For Erin and the steps still to come

ONE INTERPRETATION OF THE HUMAN FAMILY TREE

Paleoanthropologists have discovered and named over twenty-five different kinds of fossil human ancestors and extinct relatives (hominins). Throughout this book, you will meet many of these—presented here by their names and representative fossils. While we know when these hominins lived in time (shown as "Millions of Years Ago" vertically), exactly how they were related to one another remains unclear. Possible relationships are proposed by this tree, though recent fossil and genetic evidence has revealed parts of the human family tree to be an entangled snarl of interconnected branches. Future discoveries are certain, in some ways, to complicate this picture and, in other ways, to simplify it. *Illustration by Alexis Seabrook.*

Praise for *First Steps*

"A book that strides confidently across this complex terrain, laying out what we know about how walking works, who started doing it, and when. . . . DeSilva is a genial companion on this stroll through the deep origins of walking."

—*New York Times Book Review*

"DeSilva has a gift for identifying important but often overlooked observations. . . . While the subject of human evolution might seem daunting . . . DeSilva uses personable language and always keeps it interesting."

—*Library Journal* (starred review)

"DeSilva makes a solid scientific case with an expert history of human and ape evolution. . . . Accessible, valuable popular anthropology."

—*Kirkus Reviews*

"A brisk jaunt through the history of bipedalism. . . . DeSilva's ability to turn anatomical evidence into a focused tale of human evolution and his enthusiasm for research will leave readers both informed and uplifted."

—*Publishers Weekly*

"This is breezy popular science at its best, interweaving anecdotes from the field and lab with scientific findings and the occasional pop culture reference."

—*Science News*

"Before our ancestors thought symbolically, used fire, made stone tools, or even entered the open savanna, [they] walked upright. In one way or another, this odd locomotory style has underwritten the whole spectrum of our vaunted human uniquenesses, from our manual dexterity to our hairless bodies, and our large brains. In the modern world, it even influences the way other people recognize us at a distance, and it is crucial to our individual viability. In this authoritative but charmingly discursive and accessible book, Jeremy DeSilva lucidly explains how and why."

—Ian Tattersall, author of *Masters of the Planet* and *The Fossil Trail*

"Master anatomist and paleontologist Jeremy DeSilva makes no bones about the fact that when looking at fossils 'I let myself be emotional. . . .' Thus does this world expert and gifted story-teller take us on a tour through the sprawling, complicated saga of human origins. Drawing on his personal knowledge of topics ranging from sports medicine to childcare and his acquaintance with a host of colorful characters—whether lying inert in a mu-seum drawer, sitting behind microscopes, or feuding with one other—DeSilva adds flesh and projects feelings onto the bones he studies. . . . A tour de force of empathic understanding."

—Sarah Blaffer Hrdy, author of *Mother Nature* and *Mothers and Others: The Evolutionary Origins of Mutual Understanding*

"It should come as no surprise that walking matters. But what will surprise most readers is how and why. DeSilva takes us on a bril-liant, fun, and scientifically deep stroll through history, anatomy, and evolution in order to illustrate the powerful story of how a particular mode of movement helped make us one of the most wonderful, dangerous, and fascinating species on earth."

—Agustín Fuentes, professor of anthropology, Princeton University, and author of *Why We Believe: Evolution and the Human Way of Being*

FIRST
STEPS

HOW UPRIGHT WALKING
MADE US HUMAN

JEREMY DeSILVA

HARPER

NEW YORK • LONDON • TORONTO • SYDNEY

HARPER

HarperCollins books may be purchased for educational, business, or sales promotional use. For information, please email the Special Markets Department at SPsales@harpercollins.com.

FIRST HARPER PAPERBACKS EDITION PUBLISHED 2022.

Designed by Bonni Leon-Berman

Library of Congress Cataloging-in-Publication Data has been applied for.

ISBN 978-0-06-293850-3 (pbk.)

22 23 24 25 26 LSC 10 9 8 7 6 5 4 3 2 1

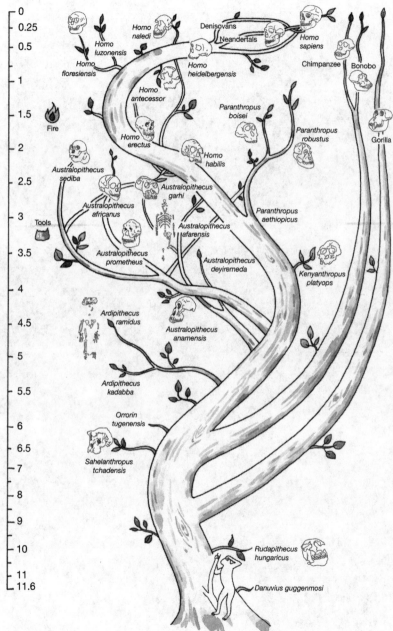

CONTENTS

Author's Note   xi

Introduction   xv

PART I: THE ORIGIN OF UPRIGHT WALKING

Chapter 1: How We Walk   3

Chapter 2: *T. rex*, the Carolina Butcher, and the First Bipeds   17

Chapter 3: "How the Human Stood Upright" and
     Other Just-So Stories About Bipedalism   31

Chapter 4: Lucy's Ancestors   47

Chapter 5: Ardi and the River Gods   67

PART II: BECOMING HUMAN

Chapter 6: Ancient Footprints   89

Chapter 7: Many Ways to Walk a Mile   113

Chapter 8: Hominins on the Move   131

Chapter 9: Migration to Middle Earth   145

PART III: WALK OF LIFE

Chapter 10: Baby Steps   163

Chapter 11: Birth and Bipedalism   179

Chapter 12: Gait Differences and What They Mean   199

Chapter 13: Myokines and the Cost of Immobility   209

Chapter 14: Why Walking Helps Us Think   221

Chapter 15: Of Ostrich Feet and Knee Replacements   235

Conclusion: The Empathetic Ape   253
Acknowledgments   269
Notes   273
Index   325

AUTHOR'S NOTE

As I write this from my home in Norwich, Vermont, a social media poll is circulating in which people report what career they have now compared to what they dreamed of doing at ages six, ten, fourteen, sixteen, and eighteen. Mine looks like this:

Age 6	Scientist
Age 10	Center fielder for the Red Sox
Age 14	Point guard for the Celtics
Age 16	Veterinarian
Age 18	Astronomer
Now	Paleoanthropologist

Paleoanthropology is the study of fossil (paleo) humans (anthropology). It is a science that asks some of the biggest and boldest questions humans have ever dared wonder about themselves and their world: Why are we here? Why are we the way we are? Where did we come from? But it was not always my path. I didn't even discover this science until the year 2000.

That year, I was working as a science educator at the Boston Museum of Science. I was making $11 an hour, George W. Bush was elected the next U.S. president, and the Red Sox had wrapped up their eighty-second season since last winning a World Series championship. My museum coworker was a brilliant science educator with the best and most contagious laugh I had ever heard. Four years later, she said yes when I asked her to marry me.

In late 2000, however, it was not love on my mind, but a terrible gaffe in the halls of the museum. Positioned within the walls of the dinosaur exhibit, *way* too close to the life-size *Tyrannosaurus rex*, was a fiberglass replica of footprints made by ancient humans 3.6 million years ago at Laetoli, Tanzania.

Like the prehistoric animal toy sets that include dinosaurs, woolly mammoths, and cavemen, positioning these footprints next to dinosaur fossils that were *twenty times* older might unwittingly promote the misconception that ancient humans and dinosaurs coexisted. Something had to be done.

I approached my boss—the great science educator Lucy Kirshner—and she agreed that the ancient human footprints should be displayed in the newly rebuilt human biology exhibit. But first she wanted me to go to the museum library and learn all I could about the Laetoli footprints and about human evolution. I devoured books on the subject and soon I was hooked. I had, as they say, caught the hominin bug. "Hominins" are what we call extinct human relatives and ancestors. The timing could not have been better. In the following two years, the oldest members of the human family tree were discovered—mysterious apelike ancestors with names like *Ardipithecus*, *Orrorin*, and *Sahelanthropus*.

In July 2002, I stood on a presentation stage at the museum with Dr. Laura MacLatchy, then a paleoanthropologist at Boston University (BU), and discussed with a fascinated public the implications of a newly discovered 7-million-year-old hominin skull in Chad, Africa. I was giddy. Here was a real paleoanthropologist talking with me about the oldest human fossil ever found.

To me, hominin fossils didn't just reveal the physical evidence for our human evolutionary history; they also contained extraordinary, personal stories of past lives. The footprints at Laetoli, for example, were a snapshot in the life of upright walking, breathing, thinking beings who laughed, cried, lived, and died. I wanted to learn how scientists squeeze information out of these ancient bones. I wanted to tell evidence-based stories about our ancestors. I wanted to be a paleoanthropologist. Just over a year after standing with Laura MacLatchy on the museum stage, I started graduate school in her paleoanthropology lab at BU and, a short time later, the University of Michigan.

Today, I teach in the anthropology department at Dartmouth College in the woods of New Hampshire and travel to Africa for my research. For nearly two decades, I have searched for fossils in caves in South Africa and throughout the ancient badlands of Uganda and Kenya. I have dug through the ancient volcanic ash at Laetoli, Tanzania, searching for more footprints made millions of years ago by our upright walking ancestors. I have followed wild chimpanzees through their jungle habitat. In African museums, I have closely examined the foot and leg fossils of extinct human relatives and ancestors. And I've wondered.

I've wondered about our large brains, sophisticated culture, and technological know-how. I've wondered why we talk. I've wondered why it takes a village to raise a child and if it has always been that way. I've wondered why childbirth is so difficult and sometimes even dangerous for women. I've wondered about human nature and how we can be virtuous one moment and violent the next. But, mostly, I've wondered why humans walk on two, rather than four, legs.

In doing so, I've realized that the many things I wonder about are all connected, and at the root of it all is the unusual way we move. Our bipedal locomotion was a gateway to many of the unique traits that make us human. It is our hallmark. Understanding these connections requires the question-driven, evidence-based approach to the natural world that I've embraced since the age of six: science.

This is the story of how upright walking made us human.

INTRODUCTION

There's an old story about a centipede who was asked which particular set of legs he used to start walking. The question took him by surprise. What had seemed a perfectly normal means of progression became a wholly perplexing problem. He could scarcely move. I'm faced with a similar difficulty when I try to account for—not how I walk, but why.

—*John Hillaby, explorer*

The year 2016 set a record for kills in the annual hunt to cull the swelling population of black bears roaming free in rural and suburban New Jersey. Of the 636 taken, 635 were dispatched with only a few howls of protest from animal lovers. But when news broke that one particular bear was dead, there was outrage.

The killing was called "an assassination." The hunter thought to be responsible received death threats. Some advocated that he, too, should be hunted and killed. Others called for his castration. Why such fury over one dead bear?

Because he walked on two legs.

Since 2014, New Jersey residents had occasionally seen the young male bear strolling down suburban streets and through backyards on two legs—a form of locomotion called bipedalism. Although he fed on all fours, an injury prevented him from putting weight on his front limbs, so to move, he reared up and walked upright.

They called him Pedals.

I never saw Pedals walk when he was alive, but as a scientist fascinated by upright walking in my own species, I wish I had. Fortunately, there are YouTube videos of him. One has over a million views; another over 4 million.

At first glance, he looked like a man in a bear suit, but once he started moving, the differences between his gait and a human's were clear. Pedals's back legs were much shorter than mine. He shuffled in quick, short steps, remaining rigid from hips to shoulders as his clawed feet skimmed the ground. It reminded me of a panicked person desperately searching for a toilet. Pedals couldn't walk upright for long before dropping down on all fours.

We are drawn to animals when they behave like us. We post videos of goats yelling like humans and Siberian huskies howling "I love you." We are amazed by crows sledding down rooftops and chimpanzees giving hugs. They remind us of our kinship with the rest of the natural world. Perhaps more than any other behavior, though, we are awestruck by bouts of bipedalism. Plenty of animals rise on two legs to scan the horizon or strike an intimidating pose, but humans are the only mammals that walk on two legs all the time. When another animal does it, we are mesmerized.

In 2011, news spread that a male silverback lowland gorilla named Ambam at the Palace of the Apes at Port Lympne Reserve occasionally walked on two legs around his Kent, England, enclosure. Soon, he was featured on CBS, NBC, and the BBC. Upright-walking-gorilla mania struck again in early 2018 when Louis, a large male gorilla, began walking around his Philadelphia Zoo enclosure on two legs because, according to many, he didn't like to get his hands dirty.

Faith the dog was born without one front limb and had the other amputated when she was seven months old. Thanks to a dedicated family that used treats to entice her to hop, she became a capable biped. She visited thousands of wounded soldiers and appeared on *Oprah*.

And in 2018, a video of a bipedal octopus circulated on social media. It used just two of its legs to propel itself along a sandy seafloor.

By our surprised reactions to upright walking in bears, dogs, gorillas, and even octopuses, we reveal how human this behavior is. When humans do it, it is ordinary. It is, you might say, pedestrian. We are the only striding bipedal mammals on Earth—and for good reason.

In the following pages, these reasons will become clear. It is a remarkable journey, which I've organized along these lines.

Part I investigates what the fossil record tells us about the origin of upright walking in the human lineage. Part II explains how it was a prerequisite for changes that define our species, from our large brains to the way we parent our children—and how those changes allowed us to expand from our ancestral African homeland to populate the Earth. Part III explores how the anatomical changes required for efficient upright walking affect the lives of humans today, from our first steps as babies to the aches and pains we experience as we age. The conclusion examines how our species managed to survive and thrive despite the many downsides of walking on two, rather than four, legs.

Come, take a walk with me.

PART I

The Origin of Upright Walking

WHY THE FAMILIAR CHIMPANZEE-TO-HUMAN IMAGERY OF BIPEDAL EVOLUTION IS WRONG

All other animals look downward; Man,
Alone, erect, can raise his face toward Heaven.

—OVID, *METAMORPHOSES*, AD 8

CHAPTER 1

How We Walk

> Walking is falling forward. Each step we take is an arrested plunge, a collapse averted, a disaster braked. In this way, to walk becomes an act of faith.
>
> —*Paul Salopek, journalist, at the start of his ten-year, 20,000-mile journey in the footsteps of our early ancestors from their African homeland to the ends of the Earth, December 2013*

Let's face it: humans are weird. Although we are mammals, we have comparatively little body hair. While other animals communicate, we talk. Other animals pant, but we sweat. We have exceptionally large brains for our body size and have developed complex cultures. But, perhaps oddest of all, humans navigate the world perched on fully extended hind limbs.

The fossil record indicates that our ancestors started walking on two legs long before they evolved other uniquely human features including large brains and language. Bipedal walking on the ground started our lineage on its unique path shortly after our apelike ancestors split from the chimpanzee lineage.

Even Plato recognized the uniqueness and the importance of upright walking, defining the human as a "two-footed, featherless animal." According to legend, Diogenes the Cynic was not pleased with Plato's description and, with a plucked chicken in hand, he disparagingly revealed "Plato's man." Plato responded by tweaking his definition of humans to include "with flat nails," but held fast to the biped part.

Bipedalism has since made its way into our words, expressions, and entertainment. Think of the many ways we describe walking: we stroll, stride, plod, traipse, amble, saunter, shuffle, tiptoe, lumber, tromp, lope, strut, and swagger. After walking all over someone, we might be asked to walk a mile in his shoes. Heroes walk on water while geniuses are walking encyclopedias. To humanize animated television characters, cartoonists draw them standing and walking on two legs. Mickey Mouse, Bugs Bunny, Goofy, Snoopy, Winnie the Pooh, SpongeBob SquarePants, and Brian the dog from *Family Guy* all walk bipedally.

In a lifetime, the average, nondisabled person will take about 150 million steps—enough to circle the Earth three times.

But what is bipedalism? And how do we do it?

Researchers often describe bipedal walking as a "controlled fall." When we lift a leg, gravity takes over and pulls us forward and down. Of course, we don't want to fall on our faces, so we catch ourselves by extending our leg forward and planting our foot on the ground. At that point, our bodies are physically lower than they were at the start of our journey, so we need to raise ourselves upward again. The calf muscles in our legs contract and raise our center of mass. We then lift the other leg, swing it forward, and fall again. As primatologist John Napier wrote in 1967, "Human walking is a unique activity during which the body, step by step, teeters on the edge of catastrophe."

The next time you look at a person from the side as he or she walks, notice how the head dips and then rises with each stride. This wavelike pattern characterizes our controlled-fall form of walking.

Of course, walking is not this clunky, and it's not this simple. To get technical for a moment, when we raise our center of mass by contracting our leg muscles, we store potential energy. When gravity takes over and pulls us forward, it converts the stored potential energy into kinetic energy, or motion. By taking advantage of gravity, we save 65 percent of the energy we would use otherwise. This ticktocking of potential energy to kinetic energy is how pendulums work. Human walking can be thought of that way—as an inverted pendulum that resembles a metronome.

Is this any different from how other animals walk when they rear up on two legs? It turns out, the answer is yes.

As a Ph.D. student, I spent a month with wild chimpanzees in Kibale Forest National Park in western Uganda. There, I met Berg. He was a large male in the Ngogo community of chimpanzees that numbered about 150—an unusually large group of apes. He was on the older side, his head hair receding a bit and his black coat flecked with patches of gray on his lower back and calves. Berg was not a high-ranking male, but occasionally he experienced a surge of testosterone, his hair puffed out, and he gave a booming pant-hoot that echoed through the forest. When he did this, it was best for humans to step out of his way.

Berg would grab a branch from the forest floor or tear one from a nearby tree, stand upright, and walk through the understory on just two legs. But he didn't move like I do. Instead, his knees and his hips were bent—the crouched kind of walk comically performed by Groucho Marx in *A Day at the Races* and other Marx Brothers films. Unable to balance on a single leg, Berg wobbled from side to side as he gracelessly crashed through the forest. It was an energetically expensive form of travel, and he tired quickly, dropping to all fours after about a dozen steps.

Humans, in contrast, are not crouched over. We stand with extended knees and hips. Our quadriceps muscles do not have to do as much work as a chimpanzee's when they walk on crouched legs.

Muscles positioned on the sides of our hips allow us to balance on a single leg without tipping over. We walk gracefully and with much more energetic efficiency than Berg did.

But why did these changes to our anatomy happen? Why did this unusual form of locomotion evolve?

Let's start our journey by considering bipedalism in the fastest human on the planet. In 2009, Jamaican sprinter Usain Bolt set the men's world record in the one-hundred-meter dash at 9.58 seconds. Between the sixty- and eighty-meter mark, he maintained a peak speed of nearly twenty-eight miles per hour for about 1.5 seconds. But by the standards of other mammals in the animal kingdom, this human speed demon is pathetically slow-footed.

Cheetahs, the fastest land mammals, exceed sixty miles per hour. Cheetahs do not typically hunt humans, but lions and leopards, who occasionally do, top out at fifty-five miles per hour. Even their prey, including zebras and antelopes, can flee snapping jaws at fifty to fifty-five miles per hour. In other words, the predator-prey arms race in Africa currently stands at no less than fifty miles per hour. That's how fast most predators run, and how fast most prey try to escape. Except for us.

Usain Bolt not only could not flee from a leopard, he couldn't catch a rabbit. The fastest among us runs at *half* the speed of an antelope. By moving on two legs rather than four, we've lost the ability to gallop, making us exceptionally slow and vulnerable.

Bipedalism also makes our gait somewhat unstable. Sometimes our graceful "controlled fall" is not controlled at all. According to the U.S. Centers for Disease Control and Prevention, more than 35,000 Americans die annually from falling—nearly the same number who die in car accidents. But when's the last time you saw a four-legged animal—a squirrel, dog, or cat—trip and fall?

Being slow and unstable seems like a recipe for extinction, especially given that our ancestors shared the landscape with the large, fast, hungry ancestors of today's lions, leopards, and hyenas. Yet here we are, so surely there must be advantages to bipedalism that outweigh the costs. The great film director Stanley Kubrick thought he knew what these were.

IN KUBRICK'S 1968 film *2001: A Space Odyssey*, a group of hairy apes gather around a watering hole on a dry African savanna. One of them looks inquisitively at a large bone lying on the ground. He picks it up, holds it like a club, and gently taps the scatter of bones around him. Strauss's 1896 *Also sprach Zarathustra, Op. 30,* begins to play. Horns: dah, dahhh, dahhhhh, DAH-DAH! Bass drum: dum-dum, dum-dum, dum-dum, dum. The ape imagines wielding the bone as a tool—a tool to kill. The furry beast rises on two legs and slams the weapon down, shattering bones and symbolically clubbing a meal, or an enemy, to death. That's how Kubrick imagined the Dawn of Man. He and his cowriter Arthur C. Clarke were dramatizing what was then a widely accepted model for human origins and the beginning of upright walking.

This model is still with us, and it is almost certainly wrong. It postulates that bipedalism evolved in a savanna environment to free the hands to carry weapons. It asserts that humans are, and always have been, violent. These ideas go all the way back to Darwin.

Charles Darwin's *On the Origin of Species* (1859) is one of the most influential books ever written. Darwin didn't *invent* evolution; naturalists had been discussing the changeability of species for decades. His great contribution was to present a testable mechanism for *how* populations changed and continue to change over time. He called this mechanism "natural selection," although most of us know it as "survival of the fittest." More than 150 years later, there's ample evidence that natural selection is a strong driver of evolutionary change.

Almost from the beginning, skeptics howled at the implication that human beings descended from apes, but in *Origin*, Darwin had written almost nothing about the evolution of his own species. He simply wrote on the penultimate page of the book that "light will be thrown on the origin of man and his history."

Nevertheless, Darwin *was* thinking about humans. Twelve years later, in *The Descent of Man* (1871), he hypothesized that humans possess several interrelated traits. He asserted we are the only apes that make tools. We know now he was wrong, but Jane Goodall's observation that chimpanzees at Gombe Stream National Park in Tanzania make and use tools was still ninety years away. However, Darwin *correctly* posited that humans are the only fully bipedal ape, and that we have unusually small canine, or fang, teeth.

To Darwin, these three human attributes—tool use, bipedalism, and small canines—were linked. As he saw it, individuals who moved on two legs could free their hands for tool use. Thanks to tools, they no longer needed large canine teeth to compete with rivals. Ultimately, he thought, this suite of changes led to an increase in brain size.

But Darwin was working with a handicap. He had no access to firsthand accounts of wild ape behavior, data that didn't start trickling in until a century later. Furthermore, in 1871 there wasn't a single known early human fossil from the African continent—the place of origin for our lineage as we understand it now, and even as Darwin predicted a century and a half ago. The only premodern human fossils known to Darwin were a few Neandertal bones from Germany misidentified by some scholars at the time as diseased *Homo sapiens*.

Without the benefit of a fossil record or accurate behavioral observations of our closest living ape relatives, Darwin did the best he could in proposing a testable scientific hypothesis for why humans walk on two legs.

Data required to test his idea started surfacing in 1924 when a young Australian professor named Raymond Dart, a brain expert at the University of the Witwatersrand in South Africa, obtained a crate of rocks from a mining operation near the town of Taung, nearly three hundred miles southwest of Johannesburg. He opened the crate and noticed that one of the rocks contained the fossilized skull of a juvenile primate. Dart used his wife's knitting needles to extract the skull from the surrounding limestone. As he did, he saw that the skull belonged to a strange primate. For one thing, the Taung child, as it would come to be known, had tiny canine teeth quite unlike those in baboons and apes. But the real clues were lurking in the child's fossilized brain.

My primary research interests are the foot and leg bones of our ancestors, but historically and aesthetically, no other fossil can match the Taung child's skull. In 2007, I traveled to Johannesburg, South Africa, to examine it. The curator there is my friend Bernhard Zipfel, a former podiatrist who became a paleoanthropologist after he "grew tired of fixing people's bunions." One morning, he retrieved a small wooden box from the vault. It was the same box Dart used to house his precious Taung nearly a century earlier. Zipfel carefully removed the fossilized brain and placed it in my hands.

After this little hominin died, the brain decayed and mud filled the skull. As millennia passed, the sediment hardened into an endocast, a replica of the brain. It faithfully duplicated the size and shape of the original brain and even preserved details of the folds, fissures, and external cranial arteries. The anatomical detail is exquisite. I carefully turned the fossil brain over to reveal a thick layer of sparkling calcite. Light reflected from it as if it were a geode, not an ancient human fossil. I hadn't expected Taung to be so beautiful.

The preservation of the folds and fissures of the brain was a remarkable stroke of luck because Dart knew brain anatomy as well as anyone in the world. He was, after all, a neuroanatomist. His studies revealed that the Taung child's brain was about the size of an adult ape's but had lobes organized more like a human's.

The endocast fit perfectly, like a puzzle piece, into the backside of Taung's skull. I turned the skull slowly to peer into this 2.5-million-year-old child's eye sockets, the closest I could come to seeing an ancient hominin eye to eye. When I rotated the skull to examine the underside, I saw what Dart had observed in 1924. The foramen magnum—the hole through which the spinal cord passes—was located directly under the skull as it is in humans. When alive, little Taung held its head atop a vertical spine.

In other words, Taung was bipedal. In 1925, Dart announced that the fossilized skull was from a species brand-new to science. He called it *Australopithecus africanus*, meaning "southern ape from Africa," following the traditional way in which scientists classify and name animals by genus and species. Domestic dogs, for instance, are all members of the same species, but they are also part of a larger group, or "genus," of related animals including wolves, coyotes, and jackals. All the members of that genus are part of a still larger and more distantly related group, or "family," that includes wild dogs, foxes, and many species of extinct wolflike carnivores.

We and our ancestors are classified in the same way. Modern humans are all members of the same species, but we are also the lone survivors of a genus that once included other humanlike groups such as Neandertals. Our genus, *Homo*, which made its first appearance about 2.5 million years ago, evolved from a species that was part of another genus, called *Australopithecus*. All members of *Homo* and *Australopithecus*, in turn, are hominids, the name for a family of related animals that includes many of the existing and extinct great apes, such as chimpanzees, bonobos, and gorillas.

Animals are referred to by their genus name followed by their species name. For example, humans are *Homo sapiens*, dogs are *Canis familiaris*, and the Taung child is *Australopithecus africanus*.

More important than the name, though, was Dart's interpretation of this fossil. He hypothesized that it was not an ancestral chimpanzee or gorilla but rather an extinct relative of humans.

While the scientific community debated the importance of the Taung discovery, another South African paleontologist, Robert Broom, searched for more *Australopithecus* fossils in caves northwest of Johannesburg in an area known today as the Cradle of Humankind. Throughout the 1930s and late 1940s, he used dynamite to blast through the hard cave walls. He then picked through the rubble, searching for the remains of our ancestors. Today, there are still large piles of cave debris—many chunks containing fossils—at the openings of these caves. They are called Broom piles.

While paleoanthropologists today cringe at his crude approach, Broom discovered dozens of fossils from two different kinds of hominins. One form, which he called *Paranthropus robustus*, had large teeth and bony attachments for enormous chewing muscles. The other, a slender form with smaller teeth and smaller chewing muscles, appeared to match Dart's *Australopithecus africanus*.

In a cave called Sterkfontein, Broom recovered a fossilized vertebral column, a pelvis, and two knee bones that demonstrated that *Australopithecus africanus* walked on two legs. We now know, from radiometric dating techniques on the uranium trapped in the limestone of the cave, that these fossils are between 2.0 and 2.6 million years old.

Meanwhile, Dart was excavating fossils at a Makapansgat cave northeast of the Cradle of Humankind. There, he discovered a small number of ancient human fossils that he regarded as different enough from his precious Taung child to be named a new species. He called the Makapansgat hominin *Australopithecus prometheus* after the Greek Titan responsible for bringing fire to humankind, because many fossilized animal bones discovered near the human fossils were charred and appeared to have been deliberately burned.

Furthermore, Dart discovered a peculiar damage pattern on the animal fossils. They had been shattered. Leg bones from large antelopes were broken in a manner that made them sharp and daggerlike. Jaws were broken in a way that one could imagine them being used as cutting tools. Dart found antelope horns that could be gripped and used as weapons. Scattered throughout the Makapansgat cave were dozens of smashed antelope and baboon skulls—seemingly the victims of a violent encounter with *Australopithecus*.

In 1949, Dart published his findings, proposing that *Australopithecus* had developed a culture that he eventually called Osteodontokeratic, combining Greek words for bone, tooth, and horn. Expanding on Darwin's ideas, he argued that the inventors of this culture used these weapons to attack other animals and one another.

Before joining the University of the Witwatersrand faculty, Dart had been a medic in the Australian army. He had spent much of 1918 in England and France, witnessing the final year of World War I. He likely had cared for soldiers with bullet wounds and burned lungs from exposure to mustard gas. Two decades later, Dart could only watch as the world around him burned again. It is no wonder that after witnessing two world wars, Dart reasoned that humans must have had violent origins and believed he had found evidence of that at Makapansgat.

Dart's ideas about human violence and the origins of upright walking were popularized by author Robert Ardrey in his 1961 international bestseller *African Genesis*. Just seven years later, Kubrick's ape-men were smashing bones to the tune of Strauss's *Also sprach Zarathustra, Op. 30*. Dart's former student Phillip Tobias was even on the set of *2001*, directing humans in ape costumes to act like a violent *Australopithecus*.

But quietly, in a laboratory in the Ditsong National Museum of Natural History in Pretoria, South Africa, Dart's ideas were unraveling.

Charles Kimberlin "Bob" Brain was a young scientist with an exquisite eye for detail. In the 1960s, he reexamined some of Dart's "tools" and found that they matched bones that had been naturally damaged or broken by the powerful jaws of leopards and hyenas. It appeared that Dart had misinterpreted these fossils. They had not been deliberately smashed by early humans.

Furthermore, the burned animal bones turned out to have been charred by a brushfire before a rainstorm washed them into the Makapansgat cave to be fossilized. Dart's *Australopithecus prometheus* was not the fire-bringer after all. Scientists also could not find enough anatomical differences between *Australopithecus prometheus* and *africanus* to justify calling them two distinct species, so *prometheus* was absorbed into *africanus*.

Meanwhile, Brain resumed excavations begun years earlier by Broom at a cave called Swartkrans in the Cradle of Humankind. There, he discovered a juvenile *Australopithecus* skull fragment that was given the catalogue name SK 54.

A few days after seeing the Taung child, I traveled to the Ditsong museum in Pretoria to study fossils from Swartkrans Cave. The collections manager, Stephany Potze, took me into the Broom Room, a small, red-carpeted space lined with glass cases that hold some of the most important human fossils ever discovered. The Broom Room has the feel of a quaint antiques shop.

There, Potze placed SK 54 in my hands. It is a thin and delicate fossil, light brown in color with occasional black patches of manganese. I was immediately struck by two circular holes about an inch apart in the back of the skull. Inside, the bone is twisted as if it had been punctured by a can opener.

Potze then handed me the lower jaw of an ancient leopard also recovered at Swartkrans.

"Go ahead," she said.

As many have done before me, I gently placed the leopard's fangs against the holes in the back of SK 54's skull. They were a perfect match.

These ancestors of ours were not the hunters. They were hunted.

In the last few decades, a host of early human fossils with bite impressions left by ancient leopards, saber-toothed cats, hyenas, and crocodiles have been discovered. A reanalysis of the Taung child found that Dart's famous discovery has talon marks in its eye sockets. A bird of prey, probably a crowned eagle, must have plucked the Taung child from the ground and carried it off to be eaten.

As so often happens in science, even the most elegant and accepted ideas wither in the face of new evidence. Even though it persists in popular culture, "Man the Hunter," who needed free hands for tools and weapons, no longer explains our bipedal origins.

Why, then, did this strange form of locomotion first evolve?

Some scholars doubt we can ever know. The fact that we are the only mammal that walks upright, it turns out, makes the mystery especially difficult to solve, but all the more fascinating.

Here's why.

Lots of animals, from sharks and trout to squid and dolphins, swim. Even extinct reptiles called ichthyosaurs swam. Yet, these animals aren't closely related at all. A dolphin is more closely related to you and me than to these other animals, and an ichthyosaur was more closely related to a falcon than to a fish. Yet, their body shapes are stunningly similar.

Why? Because it turns out that there's a "best way" to swim. The ancestors of those sharks, ichthyosaurs, and dolphins with shapes best suited for moving through water swam faster, ate more fish, and had more offspring. How could unrelated aquatic animals have such similar shapes? Because, through natural selection, a streamlined body—the best solution for moving rapidly in the water—evolved multiple times.

This is something that has happened over and over again in nature. For example, bats, birds, and butterflies all "invented" wings. Neurotoxins for poisoning prey evolved independently in snakes, scorpions, and sea anemones. Scientists call this "convergent evolution."

Can convergent evolution help us explain bipedalism? If we rely just on mammals living today, then the answer is no because we're the only mammals that do it. If other mammals regularly walked on two legs, we could study them to figure out how bipedalism helps them survive. Does it make it easier for them to gather food? Did it provide some advantage in the long-ago habitats they lived in? Could it have been some sort of mating strategy? Answering these questions on hypothetical bipedal mammals would provide some important clues for why ancient humans evolved this form of locomotion. But because there aren't any other upright walking mammals to study, weeding out the reasonable hypotheses from the outrageous ones is especially difficult.

Perhaps, then, we should peer deeper into the past—to the time of the dinosaurs. When we do so, it turns out that bipedalism is not so rare after all.

CHAPTER 2

T. rex, the Carolina Butcher, and the First Bipeds

Four legs good, two legs better! Four legs good,
two legs better! Four legs good, two legs better!

—*The sheep,* Animal Farm, *George Orwell, 1945*

As a kid, I used to watch *Land of the Lost* reruns with my brother and sister. I was always scared of the Sleestaks—reptilian creatures who spoke in hisses and were always trying to kidnap members of the Marshall family. They were very tall, had freakishly large eyes, and walked on two legs. A young Bill Laimbeer was one of the Sleestaks, which explains both their height and perhaps why I hated them. Laimbeer was six feet eleven and played professional basketball for the Detroit Pistons. I was a Boston Celtics fanatic.

Land of the Lost was, of course, made up. But bipedal reptiles were real.

As the Pyeongchang Winter Olympics were wrapping up in 2018, South Korean scientists announced the discovery of a magnificent, 120-million-year-old trail of footprints made by an upright running lizard. Perhaps to escape a predator, the lizard had reared up on two legs and sprinted across a mudflat, leaving behind impressions of bipedal locomotion. The mudflat hardened under an intense sun and was buried under years of sediment. Geological uplift and erosion eventually uncovered this moment in one ancient lizard's life, and fortunately scientists were there to discover it before the prints crumbled.

While ancient footprints are rare finds, the discovery of a bipedal lizard should not be surprising. Even today, the South American basilisk lizard can rise on two feet and run for its life. This small lizard runs so quickly that it even skips bipedally across small stretches of water. That explains the basilisk lizard's other name, the Jesus Christ lizard.

Additional discoveries make clear that speedy reptiles have been moving on two legs for quite a while. Twenty years ago, University of Toronto paleontologists discovered the skeleton of a small bipedal reptile preserved in ancient fossilized swamps of central Germany. They called it *Eudibamus cursoris*, which translates to "original runner on two legs." Its long legs and hingelike joints are clues that this extinct animal was bipedal.

Surprisingly, *Eudibamus* was old—really old. It lived about 290 million years ago, not long after the origin of reptiles themselves and tens of millions of years before the first dinosaurs evolved. At this age, *Eudibamus* is one of the earliest known animals in the history of the Earth to move on two, rather than four, legs. Although this speedy little reptile could escape from predators by dashing away on its hind legs as the Jesus Christ lizard does today, *Eudibamus* was an evolutionary dead end.

In other words, the first known terrestrial bipedal lineage was a failure. It went extinct without leaving any modern descendants. But the golden age of bipedalism was just around the corner.

OUTSIDE MY WINDOW, a little bipedal animal is skipping through a greening spring lawn searching for worms. She is adorable.

No, I'm not referring to my daughter, but to an American robin. She hops from spot to spot on the lawn, occasionally jabbing her beak into the soil for a worm. Something eventually spooks her, and she uses her feathered front limbs to fly into a nearby tree.

All birds, from robins and eagles to ostriches and penguins, are bipedal. When did birds evolve bipedalism? And why? Perhaps understanding bird bipedalism can help us understand upright walking in humans.

One approach that scientists use to understand the evolutionary origins of a particular feature is to examine the anatomy of close relatives, looking for similarities and differences. We do this all the time, comparing ourselves to our ape cousins. Who, then, are the closest living relatives to birds? The answer to this question is clear but shocking.

By studying the anatomy and DNA of birds, scientists have confirmed that our feathered friends are most closely related to crocodilians—crocodiles, alligators, and caimans. DNA and fossils can reveal not only which animals are closely related but when they last shared a common ancestor. For instance, the common ancestor of humans and chimpanzees lived a mere 6 million years ago. But the last common ancestor of all birds and crocodilians lived much farther back in time—more than 250 million years ago.

Kirk Cameron, the teen star of the 1980s comedy *Growing Pains*, appears in an online video in which he presents the bird/crocodile genetic relationship as evidence that evolution is a made-up fairy tale: "This is what evolutionists have been searching for . . . the crocoduck?"

But evolution does not work the way Cameron imagines. The last common ancestor of any two animals is not some *Island of Dr. Moreau* hybrid of two existing forms. Rather, it is usually a more generalized form from which the modern specialized animals independently changed and evolved over time as they adapted to their particular environments.

The last common ancestor of birds and crocodiles wasn't a croco-duck, but a group of animals called archosaurs. We know this because they left us fossils. Early archosaur fossils can be found in rocks dated from 245 to 270 million years ago, a time called the early Triassic. Some archosaurs ate plants. Others ate small reptiles and primitive protomammals that looked superficially like furry lizards. Some archosaurs were tiny. Some were huge. One kind, called *Postosuchus*, which easily could be mistaken by museumgoers for a miniature *Tyrannosaurus rex*, was occasionally bipedal.

Archosaurs were not dinosaurs—not yet. Around 245 million years ago, for reasons still unclear, the archosaur lineage split, and two dominant forms evolved. One eventually led to modern-day crocodiles and alligators. The other became dinosaurs and eventually birds.

At the base of both lineages stood bipeds.

IN 2015, LINDSAY Zanno of the North Carolina Museum of Natural Sciences described the remains of a Triassic crocodilian in 230-million-year-old sediments just west of Raleigh. She called it *Carnufex carolinensis*—the "Carolina Butcher." It stood nine feet tall, had a mouthful of razor-sharp teeth, and occasionally moved on two legs. We often think of crocodiles as living fossils, as though they haven't changed at all since the time of the dinosaurs. But in contrast to the sprawling, somewhat slow crocodiles of today, the earliest crocodile ancestors were lightly built, speedy animals with upright posture, and even—at times—bipedal gaits.

Zanno's paleontology lab is located on the second floor of the Nature Research Center in downtown Raleigh. Built in 2012, it is a modern facility designed to engage the public in the process of science. Behind a panel of glass, on full display to visitors, are not just fossils but the paleontologists who discover and study them. A *Triceratops* skull sits in the center of the lab, waiting to be cut from its protective plaster jacket and cleaned. The air is filled with the buzz of air scribes slowly removing ancient soil clinging to the mineralized bones of long-extinct animals.

"I always wanted to be a paleontologist," Zanno told me. In elementary school, she had dinosaur erasers that she meticulously arranged on her desktop. She had a passion for science but didn't think studying dinosaurs was a realistic career option, so, at first, she enrolled as a premed major at a community college.

That was when she read *From Lucy to Language* by Don Johanson, the discoverer of the famous early human fossil called Lucy. Zanno was hooked again on fossils but was discouraged by the field of human paleontology.

"It was clear at the time that there were more researchers than fossils," she said.

Zanno turned instead to vertebrate paleontology—to the world of dinosaurs and their relatives—and has never looked back.

She removed the *Carnufex* fossils from a cabinet for me and carefully placed them on a sheet of protective padding. The 230-million-year-old fossils are light orange. The skull fragments are thin and delicate. Zanno assembled them, revealing a head the size of an adult alligator's. Then she took out the arm bone found next to the skull. It was tiny. Having such a small arm suggested to Zanno that *Carnufex* may have occasionally moved on two legs, perhaps having the flexibility to transition between quadrupedal (four-legged) and bipedal gaits. But, as she pointed out, no leg bones have yet been recovered. One way to deal with missing information is to compare *Carnufex* to other crocodilelike archosaurs that lived around the same time. Ancient relatives of crocodiles, with names like *Popo-saurus*, *Shuvosaurus*, and *Effigia*, have been discovered in Texas and New Mexico. These ancient beasts also had large heads and small arms. The most complete fossils reveal that they had long, powerful legs. From the hingelike shape of their preserved ankles and their large heels, researchers can tell that they all moved on two legs.

The Carolina Butcher was discovered in a rock quarry just south of the town of Moncure in Chatham County, North Carolina. Mud-stone exposed to high levels of oxygen in the atmosphere during the Triassic rusted the rocks, turning them red and orange. Today, they are ripped from the ground and made into bricks, destroying any fossils preserved in the mudstone sediments. The fossils survived what the Earth has thrown at them for 230 million years. Until now. Quarry workers sometimes give scientists just days before bones are pulverized by heavy machinery. For paleontologists, then, this is a salvage operation.

I drove across the Deep River bridge on old Route 1 and entered Chatham County on a sunny March afternoon. In a seven-mile stretch, there are six churches, four of them Baptist. Daffodils were in bloom, and streets with small, single-family houses were lined with flowering dogwoods, the state flower. Many of the homes are made of Triassic mudstone bricks. Some of these bricks undoubtedly contain fossils.

But 230 million years ago, when a monster walked here, the land-scape looked very different. There were no dogwoods; flowering plants had not evolved yet. The roadside grasses hadn't evolved yet either. The plant life was dominated instead by ferns and mosses. Ancient ancestors of longleaf pine trees were joined by thick-trunked, fifty-foot cousins of today's tiny princess pines. There was no roadside trash. No cars. No people. And nowhere to pray if you happened to come across the Carolina Butcher.

A bipedal crocodile sounds like something that could have been an evolutionary success, but bipedalism ultimately was not the favored form of locomotion in the crocodile lineage. Instead, these reptiles gradually evolved quadrupedal locomotion. Occasional bipedalism may have helped the Carolina Butcher, but its croc relatives that relied more on quadrupedal locomotion were the ones that survived the test of time, perhaps because they were better suited for ambush hunting in shallow bodies of water.

Once again, bipedalism failed.

IF WE HEAD down the other path in the archosaur fork in the road, we eventually arrive at modern-day birds. At the base of this lineage stood the earliest dinosaurlike archosaurs—and the very first dinosaurs themselves. The first dinosaurs did not plod on four legs like *Stegosaurus*. They stood on their hind legs and sprinted like another later dinosaur, *Velociraptor*. They were bipedal.

As with the crocodiles, bipedalism was not a successful locomotion for some dinosaur lineages, and quadrupedalism evolved over and over again. Think of the long-necked *Brontosaurus* and the horned *Triceratops*, for example. However, one lineage of dinosaurs maintained a bipedal posture and locomotion. The fearsome *Allosaurus* patrolled the Jurassic on two legs; *Tyrannosaurus rex* ruled the Cretaceous. They were both bipedal killing machines.

Zanno doesn't just find fossils of ancient crocodiles. She spends her summers digging through Cretaceous sediments across the American West, searching for the ancestors of *T. rex* and for its bipedal cousins, called therizinosaurs. In dinosaurs, bipedalism freed the arms from the responsibilities of travel, and when that happened, extraordinary variation evolved.

My dinosaur-loving son has a sticker on his door with a picture of *T. rex*. The caption reads, "If you're happy and you know it, clap your . . . oh." The little arms of *T. rex* are the butt of many dinosaur jokes, and researchers have debated whether they served any function. But *T. rex*'s arms looked like Arnold Schwarzenegger's compared with poor *Carnotaurus*, which looked as if someone playing dinosaur Mr. Potato Head stuck a *Tyrannosaurus* head on bipedal legs, added horns, and forgot about arms entirely. Atrophy, it appears, was one option evolution had for the arms of bipedal dinosaurs.

Opening her computer screen, Zanno said, "Look at all of the amazing things that can happen to arms in bipedal dinosaurs." She showed me a reconstruction of *Alvarezsaurus*, a specialized insect-and-termite feeder whose hand bones fused together to form a pair of large digging claws. And *Deinocheirus*, a bipedal dinosaur that evolved enormous, eight-foot-long arms tipped with three long claws. Despite intimidating forelimbs that gave the species its name, which means "horrible hand" in Greek, this beast was probably herbivorous. It presumably used those claws to pull branches to its toothless beak. She also showed me *Therizinosaurus*, which, like *Deinocheirus*, had enormous arms, but this time tipped with long, flattened claws. It also had a huge belly and may have just sat while feeding. "Worst biped I've ever seen," she said.

Therizinosaurus likely possessed what many other bipedal theropods, like *Velociraptor* and *Oviraptor*, also had: feathers. Once freed from the work of locomotion, arms could be used for a variety of things. They could be used to gather food. They could be used as ornamentation to help attract a mate. *Oviraptor* likely used its feathered arms to shelter eggs in its nest. Feathers may have kept other dinosaurs warm. Over time, the feathers were co-opted for gliding and then for powered flight. Sixty-six million years ago, most dinosaurs went extinct, but some feathered, bipedal ones survived.

Dinosaurs tell us at least one thing about ourselves—bipedalism frees the arms from weight-bearing, and when that happens, innovations can evolve. In birds, of course, the arms are still used primarily for locomotion—for flight. But some birds, such as emus, ostriches, rheas, and cassowaries, don't use their arms to move. These flightless birds use their legs. Like humans, they are striding bipeds. Unlike humans, they are fast.

Ostriches can run more than forty miles per hour, while the fastest humans are lucky to sprint half that speed. These large birds have no muscles in their feet and ankles, only long tendons that stretch, store elastic energy, and recoil to propel the bird forward. Their muscles are positioned high up on their hips, like a metronome with the weight positioned far from the swinging end. With this anatomy, these birds can swing their legs very fast, roadrunner-style. While humans have longer tendons in their feet and legs than apes do, we have much more muscular feet and legs than emus and ostriches. That's why we can't swing them as rapidly.

Human and ostrich feet differ in one other important way. Human heels touch the ground; we heel strike. But the large, flightless birds walk on their toes with their heels elevated. This turns the bird foot into a spring. Some people think that bird knees bend backward. They don't. The joint we see in the same general location as the human knee is actually the elevated bird ankle, bending the same way it does in us.

Why are bird and human anatomies so different if both animals walk on two legs? Because evolution can only work with preexisting structures. We are not created from scratch. We are modified apes and, compared to the bird lineage, have been bipedal for a very short time.

Just a few million years ago, our ancestors were still climbing with mobile, muscle-filled feet well adapted for grasping tree trunks and branches. Birds are living links in an unbroken chain of upright walking animals extending back at least 245 million years. They, in an evolutionary sense, have mastered bipedalism. We are clumsy novices.

Upright walking has evolved in many different lineages, from ancient lizards and crocodiles to *T. rex* and birds. Do they have something in common that could help us solve the mystery of why we walk on two legs?

In each case, whether it's the Jesus Christ lizard or the *Velociraptor*, bipedalism appears to be all about speed. Even cockroaches stand and run on two legs when frightened. Did early humans evolve bipedalism for speed? Clearly, the answer is no. A quadrupedally running chimpanzee easily attains the same speeds as Olympic sprinters who have spent years in training. The fastest monkey, the African patas, would win Olympic gold in the hundred-meter sprint with time to spare, but it is not bipedal. We evolved bipedalism in spite of speed, not because of it.

Why does bipedalism make humans slow when it made the *Velociraptor* fast? The tail holds the answer.

Researchers from the University of Alberta found that dinosaurs' large tails contributed to their speed. The powerful tail muscles of bipedal dinosaurs attached to their back legs, giving them an added boost. Picture the posture of the clever *Velociraptors* in *Jurassic Park* or the running *T. rex* skeleton in *Night at the Museum*. The tail flexed, the head pitched forward, and off they went.

The relationship between tails and bipedalism in mammals is more complicated. Humans obviously do not have tails, and neither do any of the apes. The absence of a tail is one of the characters that defines the ape family, a group of related primates consisting of gibbons, orangutans, gorillas, chimpanzees, bonobos, and humans. The next time you are at a zoo, pause before referring to the chimpanzee, gorilla, or orangutan as a "monkey." Monkeys usually have tails, but apes do not.

All apes can move with an upright body posture, something called orthogrady, and can hang by their arms. Monkeys typically cannot do this. A more appropriate name, then, for the playground structure kids swing from would be "ape bars" rather than "monkey bars." Gibbons hang and swing under branches, orangutans move upright through the trees, and African apes climb up and down tree trunks as if they were fire poles. In humans, orthogrady, or upright-ness, is manifested in how we walk—upright bipedalism.

Clearly, one doesn't need a tail to be bipedal. Most mammals do have tails but are not bipedal. The explanation appears to be that the ancestors of mammals lost their powerful tails and evolved either small ratlike tails or, in early mammals burrowing in the shadow of the dinosaurs, none at all.

In fact, a powerful tail might actually prevent striding bipedalism in mammals. To see why, we have to go to the land down under.

SOMETIME BETWEEN 50,000 and 70,000 years ago, bands of *Homo sapiens* expanded their territory beyond the African continent and into Eurasia. They pushed eastward and eventually reached the Indonesian Archipelago. It was an ice age, and though the cold was not felt in equatorial Indonesia, its effects were. Ocean water was trapped in the bloated polar ice caps. The sea levels were low, and the "islands" of Indonesia were connected in a landmass scientists today call Sundaland. But no matter how low the sea levels were, it still took simple watercraft and a sense of curiosity and adventure for humans to keep moving southeast, skipping from island to island toward the closest large landmass—Australia.

When they landed, they found themselves in a world full of bipeds.

There were tens of millions of large, flightless, bipedally striding emus and probably even more kangaroos. But as bipeds, kangaroos don't move the way we or even emus do. They hop. For them, this is a very effective form of locomotion that takes advantage of the elastic energy stored in the long tendons of their legs. They can achieve speeds of more than forty miles per hour.

But when people first arrived in Australia, they didn't just find hopping kangaroos. They found *walking* ones. The bones of some of these animals have traveled a great distance from Australia to New York City.

The American Museum of Natural History in New York is a science enthusiast's candy shop. The lobby features an exhilarating scene in which a mama *Barosaurus* rears up on two legs to defend her baby against a hungry *Allosaurus*. The Hayden Planetarium somehow makes galactic red-shifting understandable. I could lie all day under the life-size model of the blue whale. But what few visitors realize as they explore the ring of exhibits on each floor is that the core of the museum contains vast spaces, inaccessible to the general public, where research collections are housed and science is done.

There, in April 2018, I studied the fossilized bones of gigantic Pleistocene kangaroos. The research collections are so close to the exhibits that I could hear the excited shouts of schoolkids through the walls. Cabinets stacked from floor to ceiling contain the remains of long-extinct animals. The hundreds of unopened crates containing unprepared fossils from nineteenth- and early-twentieth-century paleontology expeditions reminded me of the closing scene of *Raiders of the Lost Ark*.

Fossils too large to fit in the cabinets are housed on wooden pallets. The giant, armored shells of glyptodonts, extinct cousins of armadillos, are lined up along one wall. Next to them is an articulated skeleton of a *Glossotherium*, an extinct giant ground sloth that weighed over a ton. Against another wall is the skull of *Andrewsarchus*, an enormous carnivorous mammal that lived during the Eocene 40 million years ago. It is what I imagine the skull of the Luck Dragon from *The NeverEnding Story* would look like. Little do the visitors know that such treasures are just behind the thin walls of the exhibits.

In a far corner of this room, drawers are filled with fossil mammals from the Pleistocene of South Australia. There are two cabinets full of bones from two expeditions to Lake Callabonna, where fossils of the extinct giant kangaroo were recovered by scientists from the museum, one in 1893 and another in 1970.

The bones are dense and oddly colorful. The standard browns and grays of a fossil are joined by shades of orange, white, and even pink. Some of the bones are badly crushed. Others are beautifully preserved, still articulated together as they were in life.

From the drawers, I carefully removed the fossilized feet, legs, and pelvis of a giant short-faced kangaroo (*Sthenurus stirlingi*), which weighed over three hundred pounds and was ten feet from tail to snout. The femur alone was the size of a pipe wrench. If an animal of that size tried to hop, it would snap its tendons—a recipe for quick extinction. How did this kangaroo get around? Christine Janis, a paleontologist at Brown University, solved this puzzle. This kangaroo didn't hop. It walked.

Janis noticed that the tailbones of *Sthenurus* were relatively small, indicating that it didn't have the ability to counterbalance its body with its tail while hopping the way modern kangaroos do. Furthermore, as in humans today, the hips and knees of *Sthenurus* were disproportionately large, making them well suited for supporting this giant kangaroo's body weight on one leg at a time. Four-million-year-old footprints recently discovered in Central Australia confirm Janis's interpretation of these old bones—*Sthenurus* walked.

As I studied the fossils, I tried to imagine the sight of an enormous walking kangaroo. I wished that it were still alive, roaming the Australian outback. Unfortunately, by the Pleistocene, *Sthenurus* was gone—perhaps hunted into extinction by the new bipedal kid on the block.

THE GIANT, SHORT-FACED kangaroo couldn't hop like kangaroos today because it was too big. Other large, extinct mammals are often pictured as upright, if not bipedal, too. Museum displays of the extinct Pleistocene giant cave bear typically position them in a threatening, upright posture. Skeletons of the giant ground sloth, *Megatherium*, are often reconstructed standing upright and foraging from low-lying tree branches. While mostly quadrupedal, there is footprint evidence that *Megatherium* occasionally walked bipedally, too. And even without a single fossil recovered from below the neck, the enormous extinct Asian Pleistocene ape *Gigantopithecus* is hypothesized by some as a biped, recalling grainy videos and fanciful sightings of Bigfoot, Yeti, and Sasquatch.

So large size appears to explain why some mammals become bipedal, but does this help solve the mysterious origin of human walking? Again, the answer appears to be no. Fossils reveal that our earliest bipedal ancestors were no larger than chimpanzees.

Size is not what compelled us to stand. Neither is speed. There has to be another reason why our own ancestors started moving on two legs rather than four, but it must be different from the factors that drove upright walking in other animals. It would have to be an impetus unique to humans.

So what is it?

CHAPTER 3

"How the Human Stood Upright" and Other Just-So Stories About Bipedalism

Speculations on the origins of bipedalism are often fascinating exhibitions of ingenuity—expressing, above all, that this is a theater for intellectual daring.

—*Jonathan Kingdon, naturalist*, Lowly Origin, *2003*

Accoring to the ancient Greek writer Aristophanes, humans used to have four legs, four arms, and two faces. They were arrogant and dangerous—a clear threat to the Gods. This worried Zeus, who considered destroying the humans with lightning as he and the Olympians had done with the Titans. Instead, he came up with an ingenious plan. He split them in half. With only two legs, two arms, and one face, humans would not be nearly as threatening. Apollo patched up these split beings and knotted them at the navel. They've been wandering the Earth ever since, searching for their other half—their soul mates.

WE ARE A curious species. We seek answers to the big questions. Where did we come from? Why are we the way we are? For evidence-based answers, we turn to science. But we scientists must be careful because, in the absence of evidence, origin stories for upright walking can be just as fanciful as Zeus splitting four-legged humans down the middle. Many explanations for human bipedalism may be written in the language of science but share the same narrative arc as *How the Leopard Got His Spots*, *How the Camel Got His Hump*, and the other Rudyard Kipling *Just So Stories*.

University of Chicago anthropologist Russell Tuttle has called bipedal-origin hypotheses "scientifically informed storytelling." In the last seventy-five years, we anthropologists have offered more and more possible reasons why humans walk on two legs, publishing well over a hundred scientific papers proposing to explain the reason natural selection favored it in our ancestors.

Few have been taken seriously.

INSTEAD OF DIVING straight into why *humans* are bipedal, it might help to first consider this question from a broader perspective. Bipedal locomotion is rare in other mammals, but the ones that move on two legs the most, if only occasionally, are the primates: lemurs, monkeys, and apes. The sifaka lemurs of Madagascar skip on two legs when they come down from the trees and travel on the ground. Capuchin monkeys gather nuts and rocks in their arms and travel short distances on two legs. Baboons stand upright while wading through waist-deep water. All of the ape species, including chimpanzees and bonobos, move bipedally from time to time.

The question, then, is not how bipedal walking emerged out of nothing, but what were the conditions in which bipedal walking *frequency* increased from occasionally in other primates to all of the time in humans.

While bipedal locomotion is rare in mammals, bipedal *posture* is not. Surely, before our ancestors started walking on two legs, they had to stand. Investigating why some living mammals stand may provide insight into why early human ancestors evolved an upright posture, a prerequisite to bipedal locomotion.

Many mammals stand on two legs to survey their environment. This behavior is common, for example, in African meerkats and North American prairie dogs. Ambam, the large male gorilla at the Palace of the Apes in Kent, England, frequently stands upright to watch his keeper, Phil Ridges, prepare his meals. And when Ambam hears a sound in the distance, he stands upright, eyes fixed in that direction. These examples support the "peekaboo" hypothesis, which posits that upright posture evolved in our ancestors as a way to scan the savanna for predators.

The year Charles Darwin was born (1809), Jean-Baptiste Lamarck, a French naturalist best known for hypothesizing a largely incorrect mechanism for evolution, proposed such an idea in his book *Zoological Philosophy*. He wrote that bipedalism satisfied a "desire to command a large and distant view."

If that is correct, our ancient hominin ancestors stood to survey the grassy landscape for dangers before crouching down to avoid being spotted. This may be accurate, I suppose, but why, then, would our ancestors start *moving* on two legs? If you spot a lion and it sees you, bipedalism would be a much slower way to escape than four-legged galloping.

Others have noticed that chimpanzees and bears stand upright to warn others away from their territories or simply to sniff the air to detect their surroundings. Perhaps our ancient hominin ancestors stood to make themselves look large and threatening. Maybe that helped them survive and have more offspring. One scholar took this a step further and argued that early hominins stood to wield thorn bushes as a defense against lions.

Or maybe instead of vigilance, standing upright helped us eat.

THE GERENUK IS a lovely Eastern African antelope that stands on its hind legs to reach young, nutritious acacia leaves. Goats occasionally do this, too. In the wild, some chimpanzees stand on two legs to reach low-hanging fruit. Sometimes they do this on the ground and sometimes in the trees. Some scientists have hypothesized that our ancient hominin ancestors also stood to grab food that would otherwise have been out of reach. Those who were more upright could fill their bellies and be healthy enough to have more offspring.

Still others position our ancestors not in a grassland or at the base of fruit trees but in a watery, sedge-rich habitat similar to today's Okavango Delta in Botswana. If correct, we were swamp apes. Baboons that live today near the Okavango occasionally stand on two legs to avoid getting a faceful of water.

This wading hypothesis is a more reasonable repurposing of the "aquatic ape" idea originally pitched in the 1960s by Sir Alister Hardy and soon after by Elaine Morgan. Popularized more recently by David Attenborough in a BBC Radio show, the idea supposes that hominin bipedalism evolved in the water. It claims to explain our hairlessness, relative buoyancy, and a diving reflex present in human babies, among other cherry-picked aspects of our anatomy and physiology.

This idea was promoted in the absurdly unscientific *Mermaids: The Body Found*, which aired on Animal Planet and the Discovery Channel in 2012 and was watched by 1.9 million Americans. The program was replete with fabricated evidence and interviews by "scientists" Rodney Webster and Rebecca Davis, played by actors Jason Cope and Helen Johns.

Aquatic-ape hypothesis makes some sense until you realize a watery habitat would have teemed with crocodiles and hippos, and that we are slow swimmers. At the 2008 Olympics gold, medalist Michael Phelps swam two hundred meters in one minute and forty-three seconds. Usain Bolt ran the same distance in nineteen seconds. If you think we are slow on land, try the water. Better yet, if crocs and hippos are nearby, don't. And in 2021, researchers found that compared with other apes, humans have evolved physiological mechanisms to conserve water—the opposite one would predict if we evolved in a watery habitat.

So why did our ancestors start walking on two legs? We honestly don't know, but it troubles me to see how confident some folks are in their pet hypotheses.

In March 2019, evolutionary biologist Richard Dawkins tweeted:

Why did we evolve bipedalism? Here's my memetic theory. Temporary bipedalism is sporadic in primates: an eminently imitable trick, conspicuous demonstration of enviable skill. I think bipedal fashion meme spread culturally, sparking meme/gene (inc sexual selection) coevolution.

In other words, it started as a case of monkey see, monkey do.

I wish Dawkins had used the word "hypothesis" rather than "theory" since in science, "theory" refers to large, overarching ideas with predictive and explanatory power. Scientists use this word differently than it is used in everyday situations, where a theory is just a wild guess. Nevertheless, Dawkins thinks bipedalism started as a trend, became culturally cool, and then went viral.

Dawkins first presented his bipedalism meme idea in *The Ancestor's Tale*, a 2004 book cowritten with Yan Wong. Fifteen years later, though, Dawkins could see what his followers thought when 433 tweeted back. A handful of responses insisted that God had made humans bipedal. Others just wrote, "Huh?" But most responses were from readers who absolutely, 100 percent *knew* why humans were bipedal. Four ideas repeatedly came up. Humans became bipedal to free the hands for tools or weapons, to see over tall grass and avoid predators, to stand upright in the water, or to gain the stamina to track down prey over long distances. But the aquatic-ape proponents were especially adamant that they knew best. Science does not operate by popular vote. Just because these ideas came up over and over again does not make them right.

But everyone loves a good mystery, so let's try a lightning round of hypotheses not mentioned in Dawkins's Twitter feed:

- STEALTHY STALKING: Bipedalism allowed early hominins to sneak up on prey and hit them with rocks.
- BUM SHUFFLE: Ground-feeding apes developed a more upright posture as they shuffled from food patch to food patch on their rears.
- HOMININ FLASHERS: Females were attracted to males who stood to display their genitalia. Yep, this is a real hypothesis.
- THE ROCKY BALBOA: Our ancestors needed hands free to punch one another.
- THE TRIPPING TRAP: Quadrupedalism faded away because front and hind limbs can get tangled and lead to falls. Of course, this doesn't seem to be a problem in other four-legged animals.
- BABY IN ARMS: Early hominins migrated with large African herd animals to scavenge carcasses and needed their arms to carry children.
- DODGING PREDATORS: Bipedal hominins are better at eluding leopards and lions. Except, of course, they aren't.

- THE SCRAMBLER: Hominins evolved bipedalism scrambling up and down hills and valleys.
- MIOCENE MINI-ME: Small-bodied early hominins got to their feet to walk or run along the tops of horizontal tree branches.
- RED LIGHT DISTRICT: Males need their hands free to carry meat to females and trade it for sex.
- FIRE! FIRE!: A nearby supernova increased the frequency of forest fires that burned down ape habitats, thereby promoting bipedal locomotion.
- OSTRICH MIMICRY: Hominins mimicked bipedal ostriches to sneak up on their nests and steal eggs. I made this one up, but is it any nuttier than some of the others?

And there are many, many more. The overwhelming number of hypotheses for bipedal origins is not itself a problem. The problem is that many of them are not scientifically testable with the information we currently have.

For an idea to be scientific, it has to generate predictions that can be compared with actual data. For example, a scientist testing the Miocene Mini-Me hypothesis might predict that the earliest bipedal hominins had small bodies, were living in forested rather than grassland environments, and had adaptations for moving and eating in the trees. If the actual fossils indicate that the earliest upright walkers had large bodies, lived in grassland environments, and ate food on the ground, then the Miocene Mini-Me idea would be refuted. If the data do not match the predictions and the hypothesis is wrong, we can move on to the next idea and science can progress. If we do not frame hypotheses as testable predictions, they are just stories, as scientifically valid as Zeus shearing humans in two. A good hypothesis is a vulnerable one, and a good scientist is a nimble one able to let go of an idea should the data align against it.

So what kind of data are we looking for?

For starters, it would be useful to know *where* terrestrial bipedalism started, and it sure would help to know *when*.

Charles Darwin hypothesized in *The Descent of Man* that humans share a kinship with the African apes. Bonobos were not known to science until 1933, so Darwin only wrote of chimpanzees and gorillas, but what about the Asian apes, the orangutans and gibbons? We're a lot like them, too. How do we determine which of these apes is the most closely related to humans?

After nearly a century of debate about these relationships, scientists in the late 1960s began comparing the proteins, and eventually the DNA, of these species. The results led to a redrawing of our family tree. As Darwin had predicted, the closest living relatives to humans are indeed the African apes. Our closest are the chimpanzees and bonobos, gorillas are our second cousins, and orangutans are our third cousins.

This close relationship with chimpanzees and bonobos does not mean that humans evolved from them. They are our relatives, not our ancestors. We didn't come from them any more than they came from us. What evolutionary theory predicts is that we share a common ancestor with them. Don't make the Kirk Cameron crocoduck mistake. This common ancestor was not a humanzee or bonosapien, some blend of us and them, but rather a more generalized ape from which we both evolved.

What was this ancient ape like? When did it live? These are tough questions, but scientists have started to figure out the answers.

MOLECULAR ANTHROPOLOGIST TODD Disotell greeted me as I stepped off the cramped elevator on the fourth floor of 25 Waverly Place on the New York University campus in Greenwich

Village. He is short and fit, looking much younger than his fifty-five years. It was a cold, raw April day, but Disotell wore brightly colored shorts, canvas loafers, and a King Kong T-shirt. His short sleeves revealed tattoos of Darwin's famous "I think" line drawing of a family tree on his right forearm and Bigfoot on his left biceps. He had shaved his mohawk and instead sported a buzz cut, goatee, and orange-rimmed glasses. Disotell is one of the world's experts in anthropological genetics.

Six graduate students and postdocs looked up from their pipetting as we toured his lab, a large facility given the cramped quarters of NYU's anthropology department. Disotell proudly stated that Giorgio Tsoukalos, the *Ancient Aliens* guy, was there to witness the extraction of DNA from skulls he thought were aliens. They were not. With federal grants increasingly difficult to get these days, Disotell has taken the clever, but controversial, tack of having TV studios buy him expensive equipment in return for his expert opinion on History Channel shows that popularize fantastical tales of ancient aliens and Bigfoot.

After the tour, we headed down the street to the White Oak Tavern for lunch. Disotell ordered an apple salad with no apple, and I asked the question that brought me to Manhattan:

"When did humans and chimpanzees last share a common ancestor?"

I was expecting a deep sigh, a thoughtful sip of his drink, and a series of ambiguous on-the-other-hand-prefaced statements.

"Six million years ago," he said without hesitation. "Plus or minus half a million."

"Really?" I said. "I've seen estimates as high at twelve million and as low as five million."

"Nope," Disotell replied. "Those dates are based on flawed assumptions."

He explained that studies once counted the molecular differences in small sequences of target genes to estimate when lineages diverged. Now, however, more advanced technology allows researchers to rapidly compare tens of thousands of genes between species, including humans and the African apes. To be sure, not all will agree with his interpretation of the data, but to Disotell, the results are clear.

Much of the fossil evidence does support his conclusion that our lineage had fully split from the chimpanzee and bonobo lineage by 6 million years ago. Assuming that a new generation appears every twenty-five years or so, the last common ancestor I shared with a chimpanzee was my great, great, great-grandmother with about 240,000 "greats" in front of "grandmother." If I said "great" once every second, it would take me three full days without a break to reach the point where my lineage diverged from a chimpanzee's.

Africa is, and always has been, vast and full of diverse habitats. It would be useful to know *where* the earliest hominins lived. Were they living in the canopies of primeval forests, or venturing into the grasslands? Environmental change often brings evolutionary changes in behavior and anatomy.

Clues to a dramatic change in the African environment are hidden in ancient soils and fossilized teeth that preserve stable isotopes of carbon and oxygen. To understand this, we have to pause for a short, simple chemistry lesson.

Carbon and oxygen exist in different forms, called isotopes. Some isotopes are unstable and radioactive. These are useful for determining the absolute age of fossils, a subject we'll return to in the next chapter. The isotopes we are interested in for reconstructing ancient environments are the stable ones.

Carbon, designated ^{12}C because it usually has six protons and six neutrons, has an isotope, ^{13}C, that has one extra neutron. It turns out that when some plants breathe, they reject the carbon dioxide with the heavier isotope, ^{13}C, and preferentially incorporate ^{12}C into their tissues. These tend to be forest plants in wet, lush environments. Grasses and other plants that grow in open and more arid savannas don't mind the ^{13}C as much and absorb more of it.

When animals eat plants, they integrate carbon into their bones and teeth. The beauty of this carbon isotope is that it is *stable* and does not go away or change even after millions of years of fossilization. Scientists can grind an ancient antelope tooth into a powder, measure the ratio of ^{12}C to ^{13}C with something called a mass spectrometer, and determine whether the animal fed in a forest, a grassland, or a mixture of both.

Oxygen, or ^{16}O, also has a stable isotope, ^{18}O, with two extra neutrons. Either form can take on the role of "O" in H_2O. Because ^{16}O is the lighter of the two, water with this isotope more readily evaporates, floats upward, and forms rain clouds. When the planet is cold, the lighter oxygen falls as snow and gets trapped in polar ice. That, in turn, concentrates ^{18}O in the oceans, forming the most important global record of past temperatures that we currently have. In addition, as climates become drier and evaporation increases, African lakes and rivers also concentrate ^{18}O. The ratio of light to heavy oxygen is also detectable in fossils and provides us with a permanent record of local temperature and humidity.

Since all animals obtain water through drinking it or from the plants they eat, researchers have used the chemistry of ancient fossils to reconstruct what Africa was like millions of years ago when our ancestors began evolving. The results are intriguing.

Starting in the Miocene, at least 10 to 15 million years ago, the African continent became drier and more seasonal. Climate fluctuations became more pronounced, forests in Eastern Africa gradually became fragmented, and grasslands expanded to fill the gaps between them. Scientists now believe that terrestrial bipedalism evolved in an environment that was shifting from vast forests to open savannas with patchy clumps of forest. However, we are not sure *why* upright walking was beneficial in this new world.

One explanation is that upright posture would have helped our ancestors stay cool in the grasslands. Equatorial Africa is hot. Most animals are active at night or at dawn and dusk and compete to find shade to prevent overheating during the day. Our ancestors may not have fared well competing with carnivores and other large African mammals for these ever-shrinking spots. Those individuals who moved in a more upright manner had less of their body exposed to the sun and, at the same time, had more of their bodies exposed to breezes that would made sweating more efficient.

Biologist Peter Wheeler, who developed this ingenious, though not necessarily correct, hypothesis in the late 1980s and 1990s, also calculated that bipedal locomotion would reduce our need for water by 40 percent. I think Wheeler is right, but not about the *origins* of bipedalism. It is likely that bipedalism evolved *before* hominins moved into these open, sunny grassland environments and had to worry about overheating.

A second hypothesis has to do with energy. When humans walk a mile, we burn about fifty calories. You can recover the energy burned by walking a mile with a handful of raisins. Walking on two legs as we do is an exceptionally energy-efficient way of moving from one place to another.

Researchers from Harvard University decided to test the efficiency of walking on two legs by comparing how we walk with how chimpanzees walk. They put Hollywood chimps—those you occasionally see in Old Navy commercials—on a treadmill and measured the energy they used by attaching a snorkel-like CO_2 detector to each chimpanzee's face. Don't worry. If an ape were not happy with the situation, it would have torn the experimental apparatus from its face and perhaps pulled the researchers' arms from their sockets. Chimpanzees are very strong and temperamental.

The researchers found that chimpanzees use twice as much energy as humans when they walk, regardless of whether they move on four legs or two. There may have been times early in our evolutionary history when resources were scarce, and a group of hominins had to move across a grassy field to get from one woodland food patch to the next. Perhaps those individuals who moved on two legs rather than four used less energy and were in a better position to survive tough times. It sounds compelling. The problem, however, is that chimpanzees use more energy than humans not because they are quadrupedal but because they walk with a crouched gait. Animals that walk with a more extended knee and extended hip—whether moving on two or four legs—save fuel. Once our ancestors perfected bipedal walking on extended legs, there were energetic benefits. But at the onset, there was nothing energetically special about two-legged walking over four.

Let's circle back to Stanley Kubrick's *2001: A Space Odyssey* and the actors in ape costumes standing to liberate their hands from the duties of locomotion. Instead of freeing the hands for combat, perhaps freed hands helped our ancestors carry something much more important and fundamental for survival: food.

Louis loves tomatoes. So the keepers at the Philadelphia Zoo, the oldest in the United States, hide tomatoes throughout this western lowland gorilla's enclosure. When he is let into his yard, he knuckle-walks to familiar hiding places and collects his favorite fruit. But Louis learned the hard way what happens when a 450-pound silverback gorilla knuckle-walks with a fistful of tomatoes. They squish.

For some reason, Louis doesn't like to get his hands dirty. If it rains overnight, he will step bipedally over the damp ground to avoid muddying his knuckles. A handful of squished tomatoes is tragic to this neat-freak gorilla.

His solution? Bipedalism.

Louis gathers up tomatoes in his hands and cradles them as he walks bipedally from spot to spot in his enclosure. Michael Stern, his keeper, told me that it happens a few times a month.

Similarly, at the Palace of the Apes at Port Lympne Reserve in Kent, Ambam's sister Tamba occasionally walks bipedally when she gathers food in her arms. She does it most often when also carrying her infant.

Walking on two legs in a zoo setting is one thing, but what about in the wild?

Primatologists working in the Republic of Guinea, in West Africa, have studied a group of chimpanzees for decades. These chimpanzees are famous for using rocks to crack open the hard shells of nutritious African walnuts. But the rainforest where they live also abuts cleared land where the local people grow rice, maize, and a variety of fruits including papaya, the chimpanzees' favorite. To the frustration of the people, chimpanzees raid the farms to steal the fruit.

Susana Carvalho, an anthropologist at the University of Oxford, found that the chimpanzees are more likely to move bipedally when they collect and carry highly valued food, including papayas and African walnuts. They stuff their hands so full that they have no choice but to move on two feet. Perhaps these chimpanzees are giving us a clue to why bipedalism got started in early hominins.

This idea goes back to Gordon Hewes, an anthropologist at the University of Colorado, Denver, who proposed in 1961 that early hominins did not evolve bipedal walking to carry tools or weapons, but to carry food. He based this in part on his observation that macaque monkeys often walk on two legs when they carry food. In 1964, he reminded us of Jane Goodall's offhand remark that provisioned chimpanzees sometimes gather so many bananas in their arms that they have to move bipedally.

Owen Lovejoy of Kent State University has taken this idea even further by proposing that bipedal evolution coincided with pair-bonding in our lineage. In his model, male hominins who moved on two legs could have carried food to females. The females, in turn, may have grown fond of these generous males and formed pair-bonds. Males would no longer need large canines to threaten one another over mates. This "provisioning hypothesis" thus links canine reduction and bipedalism not around violence, as Darwin and Dart did, but around sex.

Without a living example of another mammal that has adopted this sex-based strategy, it is difficult to test this hypothesis. To many critics of this idea, it deemphasizes the role that females must have had in the evolution of bipedalism and places female hominins helpless, often up in the trees, waiting for the males—the providers—to bring home the papayas.

Although it makes sense for food to be a driving force behind the evolution of bipedalism, it would also make sense for females to play a larger role. As anthropologist Cara Wall-Scheffler told me, "The target of natural selection is a female and her infant." Unless a trait benefits the females and their offspring, there is little chance it will gain evolutionary momentum.

In the 1970s and 1980s, anthropologists Nancy Tanner and Adrienne Zihlman of the University of California, Santa Cruz, made just such a case. According to their hypothesis, early female hominins were diurnal gatherers of plants and small animals, which incidentally account for more calories than male hunting in most modern hunter-gatherer societies. Female hominins would gather more than they needed and share with other members of the group. Those individuals who walked upright could gather more food such as lizards, snails, tubers, eggs, fruits, termites, and roots.

The premium on sharing and cooperation would lead females to mate with males who were more sociable and less aggressive. These less aggressive males may also have had smaller canine teeth. In Tanner and Zihlman's hypothesis, female hominins used sticks to dig up roots and tubers and wore slings to carry their infants. In other words, technology was invented early, and by the females. It has been shown that in bonobos and chimpanzees, females are the more technologically savvy sex, so it would make sense for this to be the case in the earliest hominins as well.

Importantly, Tanner and Zihlman recognized that although carrying—whether it be food, tools, or babies—may have been the prime mover for the evolution of bipedalism, it was not the only benefit. Once bipedal, hominins could survey the landscape for enemies, intimidate and throw objects at would-be predators, and efficiently move from food patch to food patch. In other words, they argued, there is no single reason bipedalism was selectively beneficial in our ancestors. It was a package of benefits that led to hominins walking upright.

Perhaps then, searching for *the* reason for upright walking is futile. And even if we never figure out why bipedalism evolved, we can still reflect on how upright walking set into motion many of the changes to our anatomy and behavior that make us human today. But understanding the dawn of upright walking in our earliest ancestors requires us to put forward testable ideas—ones with clear predictions. We have to resist untestable narratives and do actual science. There needs to be more integration of what *could* have happened with what actually happened.

To get closer to that, we need fossils.

CHAPTER 4

Lucy's Ancestors

But we must not fall into the error of supposing that the early progenitor of the whole Simian stock, including man, was identical with, or even closely resembled, any existing ape or monkey.

—*Charles Darwin,* The Descent of Man, *1871*

People sometimes ask me when scientists are going to find the missing link between apes and humans. I tell them that we already have.

The concept of a missing link supposes that there should be evidence in the fossil record for an animal that was not human and not ape but possessed characteristics of both. In 1891, Dutch anatomist Eugène Dubois was searching for fossils along the Solo River on the Indonesian island of Java. He and his team recovered a hominin molar, the top of a skull, and a leg bone. The leg indicated that the hominin was bipedal, and the skull had a brain capacity of 915 cubic centimeters. No adult human today has a brain that small, and no ape has one that large. In fact, the skull has a brain size almost exactly halfway between that of the average chimpanzee and the average human. Voilà—a missing link.

Dubois called his discovery *Pithecanthropus erectus*, which roughly translates to upright ape-man. Today, the hominin these fossils came from is called *Homo erectus*, and paleoanthropologists have recovered dozens of them throughout Africa, Asia, and Europe. The enormity of Dubois's discovery cannot be overstated. He demonstrated that there was once a creature on this planet that bridged—at least in terms of brain size—the gap between modern apes and modern humans. The link was no longer missing.

However, the term "missing link" communicates something else—something wildly inaccurate about human evolution. It supposes that *all* of the differences between humans and apes evolved gradually and in lockstep. In other words, it implies that a hominin like *Pithecanthropus erectus*, with a brain halfway between ape and human, should also *move* as if it were half ape and half human—hunched over and shuffling inefficiently on two legs. Even Dubois thought so.

In 1900, Dubois and his son made a plaster sculpture of a naked, fleshed-out *Pithecanthropus erectus* for the Paris World Fair. The statue of the small-brained hominin looks down, expressionless, at a tool he is holding in his right hand. He is upright but stands upon apelike feet with long grasping toes, the big one sticking out to the side like a thumb.

A decade later, French paleontologist Marcellin Boule published his analysis of an almost complete Neandertal skeleton discovered in a cave in La Chapelle-aux-Saints, France. The skull is large, with a brain size bigger than the average human's. But this was no *Homo sapiens*, Boule concluded. The La Chapelle Neandertal skull has a large, projecting face and a prominent brow ridge. And it lacks the tall forehead found in most humans. But the way Boule interpreted the body was telling. He reconstructed the skeleton as hunched over and equipped with an apelike foot and grasping big toe.

The message was clear: if it was not us, it didn't walk like us. But neither of these interpretations has stood the test of time.

Additional fossils and even fossil footprints have revealed that both *Homo erectus* and Neandertals stood upright on humanlike feet. Erect, striding bipedalism goes back much farther than Dubois or Boule imagined. To see, we turn our attention to the starting point for all we know about human evolution—the fossil skeleton that is the icon of our science.

IT WAS DISCOVERED near the village of Hadar in the Afar Regional State of northern Ethiopia on November 24, 1974, by Arizona State University paleoanthropologist Don Johanson. Never before had anyone found such a complete skeleton of an *Australopithecus*, and it turned out to be from a new species of *Australopithecus* that Johanson and his colleagues called *afarensis*.

The night the skeleton was found, the discovery team celebrated and listened to the Beatles' *Sgt. Pepper's Lonely Hearts Club Band* over and over on their cassette player. Upon the umpteenth playing of "Lucy in the Sky with Diamonds," a team member suggested that they name the skeleton Lucy. It stuck.

Scientists sometimes call her "A.L. 288-1," a catalogue number that indicates she was the first fossil found at the 288th site of the Afar locality. Many Ethiopians call her Dinkinesh, an Amharic word that means "you are marvelous."

In March 2017, I traveled to Ethiopia to study Lucy's bones at the National Museum in Addis Ababa, a bustling African city nestled in hills more than 7,000 feet above sea level. The growing population is nearing that of Los Angeles. At King George VI Street, a three-story building houses a replica of Lucy that the public can visit. Behind the public museum is a large building where the real fossils are kept. This concrete fortress appears to have been built by a brutalist architect. Stairs circling a central courtyard resemble an Escher drawing.

In one wing, a basement room the size of a football field holds tens of thousands of fossils of extinct elephants, giraffes, zebras, wildebeest, warthogs, and antelopes. The buzzing of mini-jackhammers can be heard behind closed doors, concealing the slow grain-by-grain removal of hardened soil from fossil bone by patient fossil preparators.

On the third floor, a series of connecting rooms house the most precious fossils, the remains of ancient humans locked in bomb-proof safes. On the day I arrived, the windows were open. Amharic-language prayers blaring from nearby Christian churches floated into an unlit room on a warm breeze that hinted of roasted coffee beans and diesel. Power outages are a routine part of life in Addis.

Lucy was laid out in three wooden trays. One contained pieces of her skull, jaw, arms, and hand bones, another housed her ribs and vertebrae, and the third held her pelvis, legs, and foot bones. The fossils are snug in soft beds of slate-gray padding carved precisely to fit each one individually. White slips of paper label every bone from A.L. 288-1a, a skull fragment, to A.L. 288-1bz, the collarbone. Lucy's bones are tan with patches of gray and a touch of olive, the colors absorbed from minerals as her bones fossilized.

Bones tell stories, and Lucy's reveal a lot about her life. The ends of her arm and leg bones have fused growth plates, indicating that she was a full-grown adult when she died. Her wisdom teeth had erupted, but they are not very worn, indicating that she was still young. If she were a modern human, the wear would be consistent with a woman in her early twenties, but her teeth also hold clues that *Australopithecus* matured more rapidly than humans do today, making Lucy close to her late teens when she died. It is unclear how she died, though some have suggested that breaks in her bones are consistent with a fall from a tree. Two bite marks on her pelvis tell us that a scavenger got to her before she was swallowed by mud along the shores of an ancient lake.

Lucy was small, between three and a half and four feet tall. Her joints are about the size of someone who weighs sixty pounds. In other words, she was about the size of a seven-year-old modern human child. She was even small compared to other fossils of her kind found at Hadar, making it likely that she was indeed a she. Her brain volume was slightly larger than the average chimpanzee's—roughly the size of a large orange. But, unlike a chimpanzee, she walked around her world on two legs. Lucy was bipedal.

My students sometimes ask me what time period I would visit if my colleagues in the physics department invented a time machine. I don't hesitate. I would go to Ethiopia 3.18 million years ago and spend the day with Lucy. I would follow her everywhere to see how she moved, how she lived, how she may have cared for a baby, what she ate, and how she interacted with other members of her group. I would bring scientific equipment to measure every detail I could about her walk to calculate the forces she exerted on her joints.

Of course, this is not an option. Instead, we are left only with rare, fragmented bones to reconstruct the life of our ancestors. How do we *know* from just some old bones that Lucy and her kind walked on two legs? Skeletal clues can be found literally from their heads to their toes.

We don't walk with our heads, but they do reveal *how* we walk. In all animals, there is a hole in the skull—the foramen magnum—that allows the passage of the spinal cord from the brain. In animals that walk on all fours, such as cheetahs and chimpanzees, the foramen magnum is positioned toward the back of the skull, aligning the head with a horizontal spine. But in humans, the hole is at the very bottom of the skull, helping balance our head atop a vertical spine. This is the anatomical clue that led Raymond Dart to conclude that the young *Australopithecus* from Taung, South Africa, was bipedal.

Sadly, Lucy's head was found in pieces, so it is not possible to reconstruct where her foramen magnum was. However, she was not the only discovery Johanson made in Ethiopia. He found the fossilized remains of an entire group of *Australopithecus afarensis* individuals. He called them the "First Family."

One discovery at the First Family site was the back and bottom of a skull that preserves the foramen magnum. Its position is the same as in modern humans—smack at the bottom of the skull. Recent discoveries of more complete *Australopithecus* skulls have confirmed these observations. Lucy's kind held their heads upright, perched on top of vertical spines.

Although Lucy doesn't have a complete skull, she does possess a beautifully preserved vertebral column. In most mammals, the spine is horizontal or bends in a shallow crescent-shaped arc. Prior to learning how to walk, human babies also have a spine shaped this way. But once we take our first steps, the spine begins to change, developing an S-shaped curve. This curve reorients our torso and our head into a balanced position over our hips. Most important is the curve at the base of our spine—something called lumbar lordosis. It produces the small of the lower back and is unique to upright walking humans. Lucy's fossil vertebrae have this S-shaped curve, just as in people today.

From the neck down, the most obvious difference between a human skeleton and a chimpanzee's is the pelvis. A chimpanzee has a tall, flat pelvis that anchors muscles called the lesser gluteals on the back of the animal. This allows them to kick their legs backward, a motion useful for driving the body upward when climbing a tree. But when a chimpanzee walks on two legs, it is in constant danger of tipping over, forcing it to wobble from side to side—an exhausting way to walk.

The human pelvis is shorter, stouter, and bowlshaped, anchoring these same muscles on the sides of our bodies. When we take a step, these muscles contract to keep the body upright and balanced. Try this for yourself. When you walk, feel your own hip muscles tense to keep you from tipping over. They can do this because of the shape of our pelvis.

What about Lucy? Her pelvis is short and stout, like a small version of ours. Her bony hips are positioned on the side of her body, meaning that her hip muscles kept her upright and balanced when she walked around her world on two legs.

As we move south from the pelvis, we encounter one of the best places to find evidence for bipedal walking: the knee. In human newborns, the longest bone in the body, the femur, is straight. But once babies start walking, the downward pressure causes the end of the growing femur to tilt. This tilt, called a bicondylar angle, develops only in individuals who walk on two legs. Chimpanzees never develop one, nor do humans who are born quadriplegic and never take a step. If Lucy's knee had this angle, she *must* have walked on two legs because there is no other way to develop one.

Johanson found Lucy's left knee, but it was crushed and could not easily be put back together. However, the year before Lucy was discovered, Johanson visited Hadar for the first time, and the first hominin fossil he found was a knee. It had a bicondylar angle. It must have come from something that walked on two legs.

As bipeds, the only part of our body that directly touches the ground is our foot. It makes sense, then, that some of the key anatomical adaptations for upright walking can be found there. We have large heels, rigid ankles, and long Achilles tendons. Unlike every other primate on the planet, humans have a nongrasping big toe that is in line with the other toes. This, together with a long, stiff, and arched sole of the foot, helps propel us into our next step. We also have short toes that bend upward as we push off the ground. This is the opposite of an ape's toes, which are long and bend downward for grasping.

Foot bones from Lucy and the First Family are strikingly humanlike. The heels are large, and Lucy's ankle is shaped a lot like a small version of mine. While Lucy's toes were long and slightly curved, they had an upward tilt, indicating that she pushed off the ground with her toes like a human does while walking.

Lucy's bones confirmed what previous discoveries in South Africa had only hinted at: bipedalism appeared early in our evolutionary history.

The *Australopithecus* fossils discovered by Raymond Dart and Robert Broom in South African caves in the 1930s and 1940s were fragmentary—a knee here, a lower jaw there. Paleontologists had discovered only a single partial skeleton, from a young female given the catalogue name Sts 14. Her pelvis and vertebral column were also quite humanlike, indicating that she held herself upright and walked on two legs. But her head was never found.

The fossilized skulls discovered in these South African caves were of hominins with brains about the size of a gorilla's. It sure looked, then, as if upright walking came before brains started getting bigger, but before Lucy, no ancient hominin skeleton with both the head and body had been found.

Lucy was just that—an *Australopithecus* with a small, ape-size brain and a humanlike body. Dubois and Boule were wrong. A stooped-over ancestor walking in a half-human, half-ape manner and possessing a brain halfway between ape and human never existed. Instead, our large brains evolved quite late, while bipedalism appeared early.

But how early?

SEVENTEENTH-CENTURY THEOLOGIAN JAMES Ussher asserted that the Earth burst into existence at six in the afternoon on Saturday, October 22, 4004 BC, a date blessed with remarkable precision. But based on everything modern science has learned about our world, it is wildly inaccurate. Often, a trade-off between precision and accuracy exists. Geologists have opted for the latter.

Lucy is estimated to have lived and died roughly 3.18 million years ago. Compared with Ussher's calculations, this age is accurate but lacks precision. As much as I'd like to be able to write that Lucy died at 8:10 a.m. on July 11, 3,181,824 years ago, I can't. Our techniques for dating fossils give us ballpark figures rounded to the nearest 10,000 years if we are lucky. To understand why, we have to do a bit more chemistry.

When a volcano erupts, it spews molten rock and ash. Present in that toxic brew churned up from the viscous mantle of the Earth is an isotope of the element potassium (K). Most potassium on Earth is ^{39}K, but the kind we are interested in for dating fossils has an extra neutron, making it ^{40}K. It is radioactive, meaning that it is unstable and really doesn't want to be ^{40}K. As it decays, it turns into a different element, argon (^{40}Ar), an inert gas that floats harmlessly out of the rock and into the atmosphere. Lucky for us, not all of the argon escapes.

Rocks and ash created in the inferno of volcanic eruptions often contain crystals called feldspar that also include radioactive potassium. But as *this* potassium decays, the resulting argon is trapped inside the crystals. As time passes, more and more argon accumulates in them, and it does so at a constant rate known as a half-life. As the decaying potassium isotope ticks away, it provides us with a dating technique with the useful nickname "clock in a rock." Carbon dating works in much the same way but is useful only for dating objects no more than 50,000 years old. For older things, such as Lucy, we need an isotope like ^{40}K.

To understand how this works, consider a glass of beer. When a bartender pours a Guinness, there is a lot of foam. Slowly, the foam "decays" and is transformed into beer. This happens at a constant rate. Half of the foam turns into beer in a minute or so. The next half in another minute. The next half in another minute and so on, until there is just a thin film of foam left. A pint glass with a lot of foam was just poured. A glass with no foam is older. The same thing happens with ^{40}K and ^{40}Ar. A rock with a lot of radioactive potassium is relatively young. A rock with a lot of argon is older. By measuring how much of each is left, we can determine how *much* older.

Of course, Lucy's bones are not made of volcanic ash, so they are not datable. But her bones were found in sediments above a layer of hardened volcanic ash called tuff. Geologists took samples of the tuff, isolated the feldspar crystals, and used a mass spectrometer to measure the amounts of ^{40}K and ^{40}Ar inside them. The ratio resulted in an approximate date of 3.22 million years.

Since Lucy's bones were found above this layer, we know that she died no later than 3.22 million years ago, but it doesn't tell us when. It could have been 3 million years ago, a million years ago, 50,000 years ago, or in 1965. That's not very helpful.

Fortunately, Eastern Africa is tectonically active, and volcanoes erupted again, forming another tuff layer just above her bones. So she must have died sometime between those two eruptions. The date of the top ash layer is 3.18 million years.

That places her death in a 40,000-year window between 3.22 and 3.18 million years ago. Because her bones were discovered much closer to the 3.18-million-year-old tuff, we estimate that she died closer to that time. Our fossil-dating methods lack precision, but they are accurate.

When Lucy was discovered in 1974, she was the oldest partial skeleton of an extinct human ever found. Johanson's firsthand account of her discovery, *Lucy: The Beginnings of Humankind*, was a *New York Times* bestseller and the inspiration for many future paleoanthropologists. Replicas of the Lucy skeleton are a staple of science museums all over the world. Ethiopian restaurants are often named "Lucy." She was even featured in a *Far Side* cartoon. During his 2015 trip to Ethiopia, President Barack Obama made a point to visit her. At a state dinner that evening, he said:

> *We are reminded that Ethiopians, Americans, all the people of the world are part of the same human family, the same chain. And as one of the professors who was describing the artifacts correctly pointed out, so much of the hardship and conflict and sadness and violence that occurs around the world is because we forget that fact. We look at superficial differences as opposed to seeing the fundamental connection that we all share.*

Lucy is the starting point for all we think we know about human evolution.

I picked up her talus, the foot bone that connects with the shin. It is small, but solid. These are rocks, after all. Most of the organic material has long since decomposed. But these rocks preserve exquisitely detailed anatomy. The ankle joint is smooth and square—a scaled-down version of my own. I looked closely and spotted a tiny bump where a foot ligament attached and helped stabilize her foot as she walked over the rough terrain more than 3 million years ago. On her femur, the markings where her hip musculature inserted are well defined. These muscles would have contracted and kept her upright every bipedal step she took.

As the electricity in the museum snapped back on, her smooth tooth enamel glinted, and I focused on what was left of her skull. Dinkinesh indeed.

LET'S CONSIDER AGAIN Lucy's place in human history. We know from Todd Disotell's comparison of human and African ape DNA that we last shared a common ancestor with chimpanzees and bonobos roughly 6 million years ago. Lucy lived 3.18 million years ago. That means Lucy lived around the halfway point between our last common ancestor with chimpanzees and humans living today.

The discovery of Lucy was huge for our science, but it opened up a massive, almost 3-million-year gap between *Australopithecus* and our earliest ancestors. That is how science often works. A discovery answers a few questions, but it raises many new ones. What came before Lucy and her kind? What did *Australopithecus* evolve from? Did it, too, walk on two legs? How far back does bipedalism go?

For years, we had nothing to go on.

In the mid-1990s, however, Meave Leakey of the National Museums of Kenya discovered an *Australopithecus* shinbone from 4.2-million-year-old sediments at a site called Kanapoi on the west side of Lake Turkana. The fossil has a big, flat knee and a human-like ankle joint, indicating that it came from a bipedal hominin. This discovery pushed bipedalism farther into the past, but the gap between it and our common ancestor with chimpanzees was still a mammoth 2 million years.

That changed in an eighteen-month paleoanthropological frenzy from January 2001 to July 2002, when a faint light was finally shed on our earliest ancestors. Extraordinary discoveries made by three different research teams working in three different parts of Africa pushed the origins of the human lineage, and the origins of upright walking, out of the Pliocene, 2.6 to 5.3 million years ago, and into the late Miocene, 5.3 to 11.6 million years ago. These fossils revealed that upright walking goes back to the very beginning of the hominin lineage.

However, they are controversial and expose a dark side to paleoanthropology.

KENYA IS LITERALLY tearing in half. The eastern portion is part of the Somali tectonic plate, and it is moving eastward at a rate of about a quarter of an inch a year. That's twenty-five times slower than hair grows, but over several million years it has opened a deep gouge called the African Great Rift Valley.

As the earth rips apart, water pools in low spots, forming lakes. The drive from Nairobi north to the Baringo Basin, which I made in the summer of 2005, is a greatest hits tour of Kenyan rift valley lakes, passing five of the eight largest in the country: Naivasha, Elementaita, Nakuru, Bogoria, and Baringo. The shallowest, including Nakuru and Naivasha, are home to tens of thousands of flamingos, making the lake edges appear a fuzzy pink from a distance. The road is riddled with potholes, making what should be a three- or four-hour drive drag on for six or seven hours. But it is worth it.

Northwest of Lake Baringo, the torn earth exposes a tilted and fragmented layer cake of sediments spread over an area the size of New York City's five boroughs. The oldest layers go back 14 million years and preserve the fossils of ancient apes. Some of the youngest layers are only half a million years old and contain the remains of our immediate hominin predecessor, *Homo erectus*.

Somewhere in between those times, we started walking.

In late 1999, French paleontologists Brigitte Senut and Martin Pickford were prospecting in the Tugen Hills area of the Baringo Basin. There, they plucked bits of jawbone, some teeth, an upper arm, one finger bone, and several fragmentary femurs from roughly 6-million-year-old sediments. The anatomy of these hominin fossils was unlike anything known at the time, and they rapidly announced in January 2000 that they had discovered a new species that they called *Orrorin tugenensis*.

The muscle attachments on the arm and the long, curved fingers indicated that *Orrorin* had been comfortable in the trees, but the most complete femur told a more interesting tale.

At the top of the femur in all mammals is a ball that fits into a socket in the pelvis. Below the ball is a "neck" that separates the ball from the part of the bone where the all-important lesser gluteal muscles are attached. In most mammals, the neck is short, but in humans it is long, giving our lesser gluteals the leverage required to keep us balanced when we walk on two feet. With a long femoral neck, our hip muscles don't have to use as much energy when we move from one point to another.

The neck of *Orrorin*'s femur was long like ours. *Orrorin* had the ability to walk on two legs. But *did* it? If Senut and Pickford had found the *other* end of the femur—the knee part—we could be more certain, but the part they did find sure is convincing. It indicates that *Orrorin* had a body, or at least a hip joint, adapted to walk bipedally on the ground.

As compelling as they are, the *Orrorin* fossils are a reminder that the science of paleoanthropology is being done by a flawed primate. Us. Along with two decades of reasonable scientific disagreements about these fossils, they have generated drama more fitting for telenovelas than for *The Origin of Species*.

The details, which include claims of forged permits, illegal fossil collection, and Kenyan jails, can be found elsewhere, but most tragically for our science, the precise whereabouts of the *Orrorin* fossils are unknown. They are rumored to be in a safe deposit box in a Nairobi bank vault, inaccessible to the over 7 billion living relatives of *Orrorin*. The fossils—the only evidence that these ancient individuals ever existed—deserve better. They deserve to have their stories told.

Only six months after *Orrorin* was announced to the world, Ethiopian paleoanthropologist Yohannes Haile-Selassie, at the time a graduate student in Tim White's laboratory at the University of California, Berkeley, and now the director of the Institute of Human Origins at Arizona State University, announced the discovery of *Ardipithecus kadabba*, a second hominin species from the Miocene. The fossils—a jaw, some teeth, and a few bones from below the neck—were discovered in Ethiopian sediments dated between 5 million and 6 million years old.

The canine tooth was on the small side, suggesting that *Ardipithecus kadabba* was part of our lineage and not an ancient chimpanzee or gorilla. But could it walk on two legs like *Orrorin*? Maybe.

The only part of *Ardipithecus kadabba* found from the waist down was a single toe bone. It was long and curved, indicating that it came from an animal that could use its feet to grasp as apes do today. But that toe bone had an upward tilt to the base where it would have connected with the ball of the foot. This tells us that the toes of *Ardipithecus kadabba* could have bent upward, as happens when we push off the ground in bipedal walking.

Just a short time later, in 2002, the paleoanthropological community was stunned again by the announcement of another extraordinarily old and puzzling fossil. It had been found on July 19, 2001, when Djimdoumalbaye Ahouta, a graduate student from Chad, was scouring the Djurab Desert in that Central African country. He was part of a team led by Michel Brunet, a French paleontologist who had been working in those parts of Africa for years.

I spoke with Brunet at the Collège de France on Thanksgiving

Day, 2019. Fittingly, his Paris office lies in the shadow of the Panthéon. He is one of the giants of paleontology.

"I discovered the sites," he said in French-accented English. "I knew we would find something there. I said to Ahouta, 'You are the best fossil hunter. You will find it.'"

Fossil prospecting is difficult work. It is often hot, dusty, and uncomfortable. It is easy to become dehydrated as sweat quickly evaporates into the parched air. Scorpions abound. The equatorial African sun is intense and blinding. It is best to search for fossils in the morning and late afternoon, when the sun is low and casts shadows across the landscape, making it easier to catch a glimpse of a familiar shape—a femur, jaw, or skull—eroding out of ancient sediments.

Fossil hunting in Chad adds another element of danger and discomfort: land mines. Decades of fighting between the Muslim population in the north and the Christian population in the south left unexploded munitions in the Djurab Desert sands.

One morning, Ahouta came across a collection of bones scattered on the ground. The shifting dunes of the desert had recently exposed them, and the team was lucky to be there. A sandstorm could have come along at any time and reburied them. They found leg bones and jaws of antelopes, bones from ancient elephants and monkeys, and even fossils of crocodiles and fish.

Among them was a primate jaw, some teeth, and a complete but crushed and distorted primate skull.

Fossils don't come with labels. The research team had to get this skull to a museum to compare it with those from chimpanzees, gorillas, and *Australopithecus*. When they did, their conclusion was shocking.

Based on the chemistry of the surrounding rock and the composition of the animal bones found nearby, the skull was from an animal that lived between 6 million and 7 million years ago, right around the time the human lineage and the chimpanzee lineage were splitting from one another. It had a combination of anatomies no one had seen before in a fossil ape. The researchers justifiably declared it a new species: *Sahelanthropus tchadensis*. They also gave it the nickname Toumaï, which means "hope of life" in Goran. The people of the area, who speak that language, sometimes give the name Toumaï to infants born at the start of the risky and uncertain dry season.

So what is *Sahelanthropus*?

Its brain was about the size of a chimpanzee's. Its face and the back of its head were gorillalike. But unlike any African ape, *Sahelanthropus* had worn-down canine teeth, a characteristic of human ancestors. And the foramen magnum—the hole through which the spinal cord passes—was said to be positioned underneath the skull as it is in humans. If so, *Sahelanthropus* regularly held itself in an upright posture.

Does that mean Toumaï *walked* on two legs? Not necessarily. I would like to see more evidence—discoveries of foot, leg, and pelvis fossils—before I'm convinced. But the skull is tantalizing.

Sahelanthropus, found thousands of miles from both the Great Rift Valley in Eastern Africa and from the caves of South Africa, opened up a new window into our past. Maybe we've been finding fossils of early hominins only in eastern and southern Africa because those were the places we've been looking for them.

But here's where the dark side of paleoanthropology, and all the drama, come in again.

Paleoanthropologists from the University of Michigan, where I trained as a graduate student, were skeptical of Brunet's interpretations. So were Pickford and Senut, the discoverers of *Orrorin*. Together, they published a short paper questioning whether *Sahelanthropus* was a member of the human lineage. They also wondered whether the foramen magnum of Toumaï was in as humanlike a position as Brunet claimed. After all, the skull was quite crushed and distorted. Pickford and Senut proposed that Toumaï may have been an ancient gorilla.

Brunet's team responded by publishing a reconstruction of the skull generated by CT-scanning it and digitally fixing the crushed parts in the process. The result appeared to show a humanlike hole in the base of the skull, confirming their assessment that *Sahelanthropus* was upright and could have been bipedal.

This is how science is supposed to work: a legitimate challenge resulted in continued research and a deeper understanding of an ancient fossil. But then the wheels started coming off.

A foundational element of science is repeatability. In this case, that would require independent research groups repeating the Brunet group's skull reconstruction to see if they would get the same result. To do that, they would need access to the original fossil, or a high-quality replica, and/or the raw CT scans. But this has not been allowed to happen.

Instead, a full two decades after its discovery, few outside Brunet's immediate team have been allowed to see Toumaï, or even to see research-quality replicas. Even the CT scans are off-limits.

Meanwhile, it has been learned that on the day of discovery, Ahouta found more than a *Sahelanthropus* skull, jaw, and some teeth. He found a femur.

The end of the bone is broken, but the shaft would presumably contain clues about whether *Sahelanthropus* walked on two legs. In 2020, competing teams published preliminary analyses of the femur. One team argued it was from a biped; the other said it was not.

Paleoanthropologists like me are excited to hear more details about this fossil, but when I asked the seventy-nine-year-old Brunet about the femur, he shook his head back and forth and leaned across his desk toward me.

"I am a paleontologist, not a paleoanthropologist," he said. "We have found thousands of fossils and over a hundred species in Chad, but everyone wants to know about this shitty femur. Toumaï was biped. Yes? If the femur is from biped, then it is from Toumaï. Yes? If it is not from biped, then not from Toumaï."

As we talked, two research-quality casts of the Toumaï skull lay on the desk between us. Could I photograph them and take some measurements?

"No."

According to Brunet, *Sahelanthropus* was bipedal, the case is closed, and more study or additional fossils—no matter what they look like—will not change his mind.

To be sure, recovering and studying fossils takes time and money. To Brunet, it has also meant the loss of a close friend who died of malaria while they prospected together in Cameroon in 1989. When I asked him why he does not make *Sahelanthropus* available for study or allow copies to be used for science education, he again shook his head back and forth.

"I have paid too much to find this. Too much. No one tells me what to do. They wait."

THE EARLIEST PURPORTED hominin fossils are not easy to interpret. They are fragmentary and distorted. Positioned near the base of our family tree, close to the common ancestor with chimpanzees and gorillas, they naturally have a fascinating and confusing combination of human and ape anatomies. We need the collective knowledge of the entire scientific community to reveal their secrets. The more trained eyes on these fossils, the better.

With science under attack and misconceptions about human evolution widespread, we need to present to the world the physical evidence for our humble beginnings without delay. In 1938, Robert Broom opened a paper describing, among other things, a femur of an *Australopithecus* this way:

> *No apology need be given for publishing to the world at the earliest possible moment all new evidence that is discovered which seems to throw additional light on the structure of the apes that apparently are related to the ancestors of man.*

I look forward to the day that fossils of the earliest hominins—evidence of our lineage's first steps—are duplicated and available as teaching resources in every university, major museum, and K–12 school in the world.

WE CAN TRACE the origins of our lineage to Africa between 5 million and 7 million years ago. There and then, for reasons we still do not fully understand, our apelike ancestors took their first steps. But at the start of this century, the published physical evidence that these ancient apes were bipedal consisted of one crushed skull from Chad, one broken femur from Kenya, and one tiny toe bone from Ethiopia. As one researcher put it, you could toss all of the evidence for the origins of upright walking into a shopping bag and still have plenty of room for the groceries.

We needed still more fossils. Fortunately, they were on their way.

CHAPTER 5

Ardi and the River Gods

Let's just say *ramidus* had a type of locomotion unlike anything living today. . . . If you want to find something that walked like it did, you might try the bar in *Star Wars*.

—*Tim White, paleoanthropologist, 1997*

In September 1994, University of California, Berkeley, scientist Tim White and his former students Gen Suwa and Berhane Asfaw announced that they had discovered 4.4-million-year-old bones of a brand-new species of *Australopithecus* called *ramidus* in the Aramis region of the Afar Regional State of Ethiopia. *Ramid* means "root" in the Afar language, and White claimed that this species was at the very root of the human family tree, possessing the most primitive, apelike anatomies of any known *Australopithecus*.

But six months later, White, Suwa, and Asfaw published a half-page correction. The fossils they had found in the arid badlands of the Ethiopian lowlands were not from *Australopithecus*, but from an entirely new genus White called *Ardipithecus*. This correction wasn't made because White stumbled across a prehistoric birth certificate. More fossils, including a partial skeleton, had been discovered, and they showed that this ancient hominin was much more apelike than Lucy and therefore deserved its own genus and species name, *Ardipithecus ramidus*.

But White wasn't releasing any details yet.

The year *Ardipithecus* was named, I was a pimpled freshman at Cornell University. I listened to a lot of Dave Matthews Band and ate bowl after bowl of late-night ramen noodles. My passion was astronomy and all things Carl Sagan. I would not learn about *Ardipithecus* or Tim White for another few years. When I did discover paleoanthropology, I became fascinated by *Ardipithecus ramidus*. It was a potential window into Lucy's ancestors and into the origins of upright walking, and White had a heck of a lot more of its bones than had been found for *Sahelanthropus*, *Orrorin*, or *Ardipithecus kadabba*.

But the large international team White had assembled to study the fossils had gone silent. While they were meticulously excavating, cleaning, gluing, reconstructing, molding, casting, and studying the fragile fossils they had found, the rest of the paleoanthropological world waited. Some referred to it as the Manhattan Project of our science. Everyone knew that something big had been discovered, but little information was trickling out.

I started graduate school in 2003, having heard only rumors of this skeleton. I finished graduate school in 2008, having heard only rumors of this skeleton. After rediscovering White's quote about the cantina scene in *Star Wars* that opens this chapter, I foolishly but happily rewatched the 1977 classic, hoping for some information about *Ardipithecus*. In that scene, only Luke, C-3PO, a few Stormtroopers, and Greedo are shown walking, and they all walk like modern humans, which makes sense since none of them were played by an *Ardipithecus ramidus*.

Finally, in 2009, after more than fifteen years of recovery and analysis, White's team detailed *Ardipithecus* to the world in a series of papers in *Science*, the flagship journal of the American Association for the Advancement of Science. What they proposed was a complete rethinking of bipedal origins.

Hundreds of fossils of *Ardipithecus ramidus* had been discovered, but the gem in the collection was a partial skeleton nicknamed "Ardi." Based on her relatively small canine teeth, Ardi was most likely an adult female. She lived and died in a roughly 100,000-year window between 4.385 million and 4.487 million years ago, an age determined by Ethiopian-born geologist Giday WoldeGabriel using the volcanic ash layers sandwiching her bones.

That meant Ardi lived more than a million years before Lucy.

At that time in Africa, the grasslands were expanding and the forests were retreating. Surprisingly, though, Ardi's bones were found alongside forest-dwelling animals and seeds of forest trees and plants. Evidence from carbon and oxygen isotopes indicated that Ardi lived and died in a wooded environment.

Their study of her bones, White and his colleagues concluded, told them that she was at least occasionally bipedal. According to them, this meant all hypotheses that bipedal origins began in a grassland—from standing to see over tall grass to walking upright to keep the body cool—had to be wrong. Ardi, they concluded, tells us that upright walking started in the woods.

But how can we be sure that Ardi walked on two legs—or even what her place is on the human family tree?

At 4.4 million years old, Ardi might have lived too late in time to represent the origin of our lineage, which was closer to 6 million years ago. Furthermore, as Jonathan Kingdon wrote in his book *Lowly Origin*, "We do not even know whether it is more useful or truthful to see *Ardipithecus* as the last of an older kind or the first of a newer kind."

Ardi's bones show that she was comfortable in the trees. She had long arms and long, curved fingers, and her apelike big toe stuck out of her foot like a grasping thumb. But when she came down to the ground, she didn't knuckle-walk like a chimpanzee or a gorilla.

Ardi's hand and wrist bones don't have any of the features found in a knuckle-walking ape. Furthermore, as with humans and Lucy, Ardi's pelvis shape would have allowed her to stay balanced as she moved around on two legs.

I traveled to Ethiopia in 2017 to see Ardi's foot. Ardi's bones are a light peach color. They are delicate and not as dense as Lucy's fossilized remains. In the field, White's team infused the brittle bones with a glue that kept them from crumbling like chalk as they eroded out of the ancient Ethiopian hillsides.

I studied each of Ardi's foot bones under the watchful eye of Ethiopian paleoanthropologist Berhane Asfaw, codirector with White of the Middle Awash project. Their group has been making important archaeological and paleontological discoveries in Ethiopia since 1981. Dartmouth College graduate student Ellie McNutt was there with me as was my friend from South Africa, foot expert Bernhard Zipfel. Together, we carefully examined each of Ardi's foot bones to assess White's claim that she was a biped.

Photographs and 3D surface scans were off-limits, in part because White's team had not concluded their own study of the fossils. But in those bones, I saw what White and his colleagues had. Ardi had some of the key anatomies that humans have for walking bipedally. However, she definitely didn't walk like we do.

Ardi's ankle looks a lot like an ape's. Her foot would not have naturally rested flat on the ground like a human's, but instead would have been more mobile, allowing her to grasp onto a tree trunk when she lifted her leg. The inside of her foot looks like a chimpanzee's, complete with a powerful grasping big toe, but the outside of her foot looks more like ours. Her bones would have locked together to form a stiff, rigid platform useful for pushing against the ground as she moved from place to place on two legs.

The bones of *Ardipithecus ramidus* tell a remarkable story about our own feet. They evolved from the outside in. The human foot is a fascinating mosaic of anatomies that were patched together over millions of years. The outside is ancient, having achieved a human-like form early in our evolutionary history, certainly by the time of *Ardipithecus* and probably earlier. But the inside changed more recently—by the time of Lucy. Our toes show perhaps the most recent evolutionary change, having shortened and straightened in just the last 2 million years.

Ardi's foot bones indicate that the earliest bipedal hominins pushed off the outside of their feet as they walked on two legs. They did not transfer weight to their big toes, as most of us do today, because they stuck them out to the side like thumbs, poised to grab on to tree branches. Walking bipedally with a foot like this was probably not efficient, but for *Ardipithecus* it was good enough. It was a compromise for an animal that spent a lot of time climbing in the trees but came to the ground to move from food patch to food patch on two legs.

Finding an early biped that combined upright walking and climbing should not be surprising. We already had evidence for such an animal in 6-million-year-old fossils from *Orrorin* and in the other form of *Ardipithecus*, called *kadabba*. What is surprising is the way White and his longtime colleague Owen Lovejoy used *Ardipithecus* to craft a new way to think about the origins of bipedalism. It was nothing short of revolutionary.

ONE OF MY favorite books as a kid was the *Giant Golden Book of Dinosaurs and Other Prehistoric Reptiles*. Several decades have passed since I last flipped its pages, but I still remember the vivid scenes. I can picture an enormous *Brontosaurus* in a swamp, its mouth full of leaves. I remember being simultaneously scared and fascinated by the image of a long-necked dinosaur being attacked by an *Allosaurus*.

These gripping images were drawn by Rudolph Zallinger, a Russian-born painter best known for his 110-foot *Age of Reptiles* mural at Yale's Peabody Museum of Natural History. The mural formed the basis for the illustrations in the Golden Book. In 1965, Time Life published a twenty-five-volume set of science books. There were books on the planets, the ocean, insects, the universe, and the different continents. They were beautifully illustrated, and Zallinger was commissioned to produce artwork for one book in particular: *Early Man*.

He skillfully arranged artistic renditions of known ape and early human ancestors across a four-page folding insert. Our crouched predecessors slowly and surely rose from left to right across the pages. At first, this transition was reluctant, as our ancestors maintained a crouched posture. But eventually, around the time of Cro-Magnon, they assumed a fully upright human stance. This image, called the March of Progress, was eventually co-opted into the familiar, misleading icon that appears on coffee mugs, T-shirts, and bumper stickers.

Google "human evolution" and you are certain to see image after image of a chimpanzee slowly turning into a human. Sometimes the images are silhouettes. Sometimes they picture a dark chimpanzee slowly turning into a white-skinned man who looks uncannily like Chuck Norris—imagery that reeks of racism and sexism. Sometimes there is a red line through this sequence—a creationist protest.

This iconic image is what many people think of when they hear "human evolution." It communicates simply and clearly that we humans evolved in a linear fashion from a knuckle-walking chimpanzee. One problem, though: it is wrong. As we've seen, chimpanzees are our cousins, not our ancestors. It is unlikely that they have remained unchanged for 6 million years. Furthermore, this imagery suggests that as humans evolved, our uprightness, brain enlargement, and hairlessness all appeared in lockstep. But that is not what happened. The changes that occurred as humans evolved unfolded at different rates, and some changes happened earlier than others.

In Zallinger's defense, he never suggested that humans evolved directly from chimpanzees, who are nowhere in his March of Progress. Yet implicit in his artwork is the hypothesis that human ancestors passed through a knuckle-walking phase and that the earliest bipeds crouched over like upright chimpanzees. This is a reasonable, scientifically testable idea. Since we are most closely related to chimpanzees, bonobos, and gorillas—all knuckle-walking apes—it makes sense that the common ancestor would also be a knuckle-walker.

If the last common ancestor was *not* a knuckle-walker, chimpanzees and gorillas must have evolved this form of locomotion independently. Many experts have regarded this as unlikely, but not Owen Lovejoy and Tim White. According to them, the *Ardipithecus* skeleton is definitive evidence that our ancestors never did walk on their knuckles. In fact, in their view, the human skeleton may be more primitive—and the great apes more evolved—than any of us thought.

Lovejoy and White's idea flipped the narrative of human evolution on its head. They argued that the living apes' bodies were too specialized to provide the raw material for bipedal walking. How do you get from an ape to a human? As we would say in northern New England, "You can't get there from here."

But if bipedal walking hominins did not arise from a chimpanzee-like knuckle-walker, what *did* we evolve from? Lovejoy and White proposed that African apes and hominins branched separately from something like a big, tailless monkey. As they saw it, the key was that monkeys and many primitive apes, like humans, have a long lower back and can pull their torsos over their hips and stand up-right like a person in a monkey costume. The big living apes, in contrast, have short and stiff lower backs, making it more effective for them to climb high in the trees but preventing them from standing upright without bending their knees and hips. Lovejoy and White claimed that when *Ardipithecus* stood on two legs to move bipedally, it would not have crouched over. It would have stood upright like you and me, though the grasping big toe wouldn't have let it stride like we do today.

If they are right, at no point did our ancestors walk on their knuckles or with a hunched gait.

But *Ardipithecus* lived about 4.5 million years ago, still almost 2 million years *after* the common ancestor we shared with chimpanzees. Can we bookend Ardi with something older? Something that lived 7 million to 12 million years ago? For that, we have to go back in time to the Miocene. And, surprisingly, we have to leave Africa.

APES HAD EVOLVED in Africa by 20 million years ago. We know this because genetic evidence pinpoints the last common ancestor of the living apes to that time and because the oldest ape fossils from sites in Kenya and Uganda are about this age. Apes are few in number today, but they were plentiful on the landscape in the past. They diversified into many different kinds that paleontologists have given names such as *Kamoyapithecus, Morotopithecus, Afropithecus, Proconsul, Ekembo, Nacholapithecus, Equatorius, Kenyapithecus,* and many more. They were similar to modern apes in certain ways. For instance, they didn't have tails, and there is evidence in their teeth that they ate fruit and had long childhoods. But most of them couldn't hang from their arms under branches like modern apes do and instead moved around on all fours like big tailless monkeys.

Starting around 15 million years ago, however, apes became rarer and rarer in Africa. Instead, ape fossils from that period turn up in Saudi Arabia, Turkey, Hungary, Germany, Greece, Italy, France, and eventually Spain. The great forests of equatorial Africa had shifted north, hugging the Mediterranean. There, they provided a rich environment for the fruit-eating ancestors of the modern great apes, including us. The ancient apes of Europe diversified, and again there is a laundry list of names: *Dryopithecus*, *Pierolapithecus*, *Anoiapithecus*, *Rudapithecus*, *Hispanopithecus*, *Ouranopithecus*, and everyone's favorite—*Oreopithecus*, the cookie ape.

It is strange to think of apes in Europe. Although it was warmer and wetter there at that time in the Earth's history, the tilt of the planet made these northern forests seasonal. During the dark months of winter, fruit was limited—a tough situation for fruit-reliant apes. How did they survive? Genetics, along with a simple lesson in body chemistry, may hold the answer.

Uric acid is a normal metabolic byproduct, formed when our cells break down certain chemical compounds. It is removed from our bodies when we pee. Most animals, including most primates, also get rid of it by making an enzyme called uricase that helps break it down when it accumulates in the blood. But not us. Humans have the gene to make uricase, but it is broken—a mutation stops us from making it. This is also true for chimpanzees, bonobos, gorillas, and orangutans, a fact that allows molecular geneticists to time the mutation of this gene to when we last shared an ancestor with these great apes around 15 million years ago.

It is possible that this mutation conferred no benefits on our predecessors. The problem with that explanation is that not making uricase predisposes us to gout, which causes painful arthritis at the base of the big toe. It is thus unlikely that this mutation would have been maintained unless it was also somehow beneficial.

What was the benefit? Uric acid helps convert fructose, the sugar found in fruit, into fat. Storing fat would have been useful for apes living in seasonal forests, when low levels of light in the winter months could lead to periods of starvation. No such problem exists in the equatorial rainforests of Africa, so an evolutionary change to solve it would have arisen only in the temperate forests of southern Europe.

Even with fat reserves, hungry apes would have been desperate to eat whatever they could in the winter. Today, during difficult times in their Southeast Asian forests, orangutans fall back on tree bark or unripe fruit. In contrast, genetic evidence indicates that our ancestors developed a taste for overripe, fermented fruits.

The three most consumed beverages in the world are water, tea, and beer. But unlike the first two, beer is rich in calories. A can of my favorite beer, Lawson's Sip of Sunshine, has the same number of calories as a McDonald's hamburger. Fermented fruits are calorie-rich, but only if your body can metabolize the ethanol. Otherwise, it is toxic. Most humans have a gene that produces an enzyme essential for breaking down ethyl alcohol. Chimpanzees, bonobos, and gorillas have it, too, but orangutans do not. Neither do any of the other primates except for the peculiar aye-aye of Madagascar. The presence of this gene in African apes and humans suggests that our last common ancestor relied on old, fermented fruits collected on the forest floor for calories during difficult times.

As the world cooled and dried in the late Miocene, the temperate forests around the Mediterranean could no longer support apes, and they eventually went extinct, but not before giving rise to the last common ancestor of humans and the African apes. The edges of the forests retreated south toward Africa, and with them came the arrival of the ancestors of chimpanzees, gorillas, and humans.

What did these ancient apes look like, and how did they move?

To understand, I traveled to Germany to meet Madelaine Böhme, a paleontologist at the University of Tübingen.

As a young girl growing up in Plovdiv, Bulgaria, Böhme used to sneak into archaeological sites and find Bronze Age artifacts in the scrap piles. She was fascinated by the past, and everywhere she went, she dug. By her late teens, she had unearthed a rare fossil elephant jaw from a Bulgarian hillside. Her father kept a large garden and begged her to apply her digging skills to the family plot, but she was always more interested in bones and artifacts than vegetables.

Böhme trained as a geologist and paleontologist and developed an expertise in the Miocene of central Europe. Ten million to fifteen million years ago, the swamps, forests, and river channels there teemed with ancient turtles, lizards, otters, beavers, elephants, and rhinoceroses. Fearsome cats and hyenas prowled these lands as well. We know this, in part, because of the 11.62-million-year-old fossils Böhme has discovered at a clay quarry called Hammerschmiede on the outskirts of Pforzen, a tiny town in the shadow of the Alps in the Bavarian region of southern Germany.

"Eighty percent of the fossils at Hammerschmiede are turtles, but every fossil matters," she said with contagious enthusiasm when I visited her lab. "I collect every bit."

Böhme's approach to collecting fossils contrasts with the standard paleontological practice of leaving seemingly unimportant fragments and unidentifiable bits in the ground or in discard piles. She is a paleontological vacuum. She has to be. Hammerschmiede is an active clay quarry. The landowners permit Böhme to collect there, but they also have an agreement with a mining conglomerate that extracts the thick clay deposits sandwiching the sand channel where the fossils are embedded. The excavator does not discriminate, and flecks of blackened Miocene bone are surely embedded within the rectangular clay bricks produced from Hammerschmiede, not unlike the situation I encountered with the Carolina Butcher in Chatham County, North Carolina.

Everything changed for Böhme on May 17, 2016, when her student Jochen Fuss unearthed the upper jaw and partial face of an ape. Knowing the mining company planned to rip clay from this area soon, Böhme worked quickly, shaving away the vertical sediment with her rock hammer. A lower jaw appeared—a perfect match for the upper.

Soon after, Böhme noticed a small bit of round, blackened bone and figured it was a piece of yet another turtle. But since every fossil matters, she collected it. She infused it with some glue and began excavating the rest of it.

Flicking away the clay and sand, she discovered that the bone, still partially buried in the hillside, could not possibly be from a turtle. The next guess was that it was a piece of horn attached to a bit of skull from a hoofed mammal called a *Miotragocerus*, a common animal from the Miocene of Europe whose living relatives, called blue bulls, are native to India. She handed her tools to Fuss to finish the excavation, but when he reached the end of the bone, it didn't taper to a point as expected. Instead, it flared outward.

"That's impossible," Böhme said.

Impossible for an antelope horn, but this was no horn. They had unearthed the ulna, or lower arm bone, of an ape. It was extremely long, comparable in length to the arms of apes able to hang from tree branches today. Additional chunks of blackened bone were infused with glue and packaged in plaster for preparation and study in the lab.

"This is where we found it," Böhme told me on a blustery day in November 2019.

Hammerschmiede looked less like a fossil site than a gravel pit. The quarry was shaped like an amphitheater, the center low and stripped clean. Thick deposits of gray clay rose around us, the rim of the quarry lined with evergreens. Böhme pointed out the lens of sand where the fossils are preserved, but I was distracted by the excavators and bulldozers that appeared poised to start work at a moment's notice.

Two weeks earlier, Böhme's team had published their analysis of the fossil ape bones from this quarry, announcing a new species of Miocene ape called *Danuvius guggenmosi* that had lived there more than 11 million years ago. Danuvius was a Celtic-Roman river god for whom the nearby Danube was named.

But I wasn't there because a new fossil ape had been discovered in Germany. I was there because Böhme claimed *Danuvius* had moved on two legs.

In 2017, after it became clear that its face and teeth distinguished it from any other known Miocene ape from Europe, Böhme was already writing the manuscript describing *Danuvius guggenmosi* in her office. In the lab next door, Thomas Lechner, a master's degree student in geology and paleoecology at the University of Tübingen, was carefully removing clay and sand from the fossils that had been rapidly collected from the quarry the year before. One of them, generically identified in the field as a "mammal long-bone," turned out to be an ape tibia, or shinbone.

I had traveled almost 4,000 miles to see it.

Böhme's office is on the second floor of the University of Tübingen paleontology museum. Down the hall in one direction are spectacularly preserved fossil ichthyosaurs, ammonites, and ancient sea lilies recovered in Jurassic shale quarries in the nearby town of Holzmaden. In the other direction are fossil dinosaurs and therapsids, the furry, reptilian ancestors of mammals.

I sat at a round table near cabinets full of fossils, unsure if I would be able to see or measure any of them. I didn't know Böhme, and she didn't know me. I've learned, sometimes the hard way, that some discoverers of ancient human and ape fossils restrict access to a limited few in bizarre quid pro quos antithetical to how science should work. I had nothing to offer Böhme except my knowledge of the shinbone, the subject of my Ph.D. thesis.

Within minutes, though, I was surrounded by fossils from Hammerschmiede. Böhme enthusiastically placed the *Danuvius* ulna in front of me, followed by a toe bone, finger bones, femurs, and finally the tibia. Sprinkled in were a baby elephant pelvis, monkey fossils from a younger site in Bulgaria, and replicas of fossil apes from Pakistan, Spain, and Kenya. Böhme loves fossils as much as I do, and she agrees with my views on our science—these bones are meant to be shared and studied, not hidden away.

I pulled out my calipers and camera and got to work.

The *Danuvius* tibia is complete, providing insight into the function of both the knee and the ankle joints in this ancient ape. The way the knee joints of all animals flex and extend depends on how the rounded, bony end of the femur rolls over the top of the tibia. In the apes, the top of the tibia is rounded, giving chimpanzees, gorillas, and orangutans much more knee mobility than we have. Surprisingly, the knee of *Danuvius* is flat like ours, and that would have allowed it to extend its legs and stand more upright.

What shocked me most, however, was the ankle joint.

In all living primates except humans, the very bottom of the tibia, where the leg meets the foot, is tilted. This orientation twists the feet inward so that they are ready for grasping. It also angles the tibia so that the knees are positioned far from one another, making apes bowlegged. But in humans, the ankle joint is flat, positioning the knees next to one another and directly over the feet.

The *Danuvius* ankle was similar to ours, or better yet, Lucy's. Furthermore, the shape of two backbones also found at Hammerschmiede convinced Böhme that *Danuvius* had an S-curved spine, a crucial anatomy for erect posture in humans and our bipedal ancestors. Böhme and her team concluded that over 11 million years ago *Danuvius* was upright and walked not on the ground, but in the trees. If Böhme is right, bipedalism did not emerge from the ground up, but from the trees down. From my own observations of the fossils, I saw no reason to contradict these findings, but they remain controversial and contested.

Böhme's hypothesis, it turns out, was foreshadowed nearly a hundred years ago.

IN 1924, LONG before genetic evidence had positioned humans in the ape family tree next to knuckle-walking chimpanzees and bonobos, Columbia University surgeon and foot expert Dudley J. Morton predicted an ape like *Danuvius*. Morton's specialty was podiatry, but he was also interested in evolution. He proposed that the best model for understanding the evolution of bipedalism in humans was the most bipedal of the apes: the gibbon.

Gibbons today are brachiating specialists, using their comically long arms and hands to rapidly swing from branch to branch in Southeast Asian tropical forests. Their arms are so long that a standing gibbon can place its hands flat on the ground without bending over. The problem, however, is that their arms are so long and thin that a gibbon cannot put too much of their body weight on them without risking a fracture. So what do gibbons do on the ground? They raise their arms in the air and run on two legs. Even in the trees, a gibbon will sometimes walk along the tops of branches and use its arms for balance, like a tightrope walker.

With no fossils, and no knowledge of molecular genetics, Morton worked with what he had to compare the bones and behavior of different ape species. He concluded that the ancestor of humans must have looked like a big gibbon, but with shorter arms. He imagined it to have powerfully grasping hands and feet and to move bipedally in the trees.

His ideas remained influential throughout the mid-twentieth century but fell out of favor by the late 1960s. First protein comparisons and then DNA studies showed that of all the apes, gibbons were the *least* related to us. Because we were more closely related to chimpanzees and gorillas, it was easier to imagine our ancestors as large, terrestrial knuckle-walkers.

The problem is those pesky fossils.

Not a single one has been found from a large, knuckle-walking ape that lived when humans, chimpanzees, and gorillas split from their common ancestor. Instead, what few fossils we do have from this time period are from smaller apes that can hang by their arms from tree branches and have flexible backs capable of upright posture.

Just weeks before Böhme introduced *Danuvius* to the world, University of Missouri paleoanthropologist Carol Ward published her study of a pelvis of *Rudapithecus hungaricus*, a 10-million-year-old fossil ape found in the swamp deposits of Rudabánya, Hungary. Its skull and teeth position this animal at the base of the great ape lineage, but the pelvis doesn't look great ape–like at all. Instead, it is similar in many ways to a siamang, the largest of the gibbon species. The pelvis indicated to Ward and her colleagues that *Rudapithecus* could move upright more effectively than any modern great ape. It turns out that *Danuvius* may not have been the only ape walking in the trees.

"Asking why humans stood up from all fours is the wrong question," Ward told my human evolution class when she visited Dartmouth College in 2018. "Perhaps we should instead be asking why our ancestors never dropped down on all fours in the first place."

BACK AT HAMMERSCHMIEDE on that same trip, I slowly spun in place, taking in the full panorama of the quarry. The place where the large male *Danuvius* skeleton had been found three years earlier was gone—stripped clean by excavators. So much earth had been removed—so much of *Danuvius* potentially turned into bricks.

"We found the female *Danuvius* right there," Böhme said, pointing to a wall of clay sandwiching a lens of sand where her team had recovered a few teeth and a small femur. Ever the optimist, she smiled and said, "I'm certain there is more of her in there. We'll find her next year."

DANUVIUS IS WONDERFULLY disruptive for our science. Additionally, it might help us solve the mystery of other controversial fossils.

Recall that Toumaï, the *Sahelanthropus*, lived and died 7 million years ago. Some molecular geneticists have argued that the split between the human and chimpanzee lineages occurred more than 7 million years ago, which would allow a bipedal *Sahelanthropus* to just eke by as a hominin. Others insist that the last common ancestor lived more recently and that *Sahelanthropus* has been misinterpreted as a bipedal hominin.

Danuvius might provide a resolution to this dilemma.

Todd Disotell, our tattooed molecular anthropologist friend from Chapter 3, firmly places the last common ancestor at 6 million years ago, plus or minus 500,000 years. If he is right, Toumaï, an upright, potentially bipedal ape, predated the last common ancestor. It would have existed *before* hominins evolved—before there was a human lineage at all. This couldn't be true if upright walking is unique to hominins, but it makes sense if the last common ancestor of humans and the African great apes was more bipedal than chimpanzees and gorillas are today.

That's where *Danuvius* comes in.

Danuvius may be telling us that our ancient ape ancestors walked upright in the trees, holding their hands above their heads to grip branches as they moved along narrow tree limbs to reach ripe fruit. Orangutans sometimes do this today, as do gibbons and spider monkeys.

From an ancestor like these, the predecessors of gorillas and, soon after, chimpanzees may have split off, evolving larger body sizes that made falls from trees more deadly. They evolved longer arms, larger palms, and stiffer lower backs to keep themselves from fatal plunges to the forest floor.

But in African forests, fruit-bearing trees are patchy, and the canopy cannot be easily bridged, so long-armed, big-bodied apes would have needed to travel over ground to get to the next food patch. Could they have moved on two legs? No. Their short, stiff lower backs would have forced them into a crouch, making travel on two legs exhausting. They would have had to move on all fours, but their fingers were too long and curved to place flat on the ground. They would have curled them under the knuckles. If this scenario is right, chimpanzees and gorillas evolved knuckle-walking separately, but for the same reasons.

And our ancestors? They were already adapted for an upright posture, and, with a few anatomical tweaks, bipedal walking on the ground could be as simple as bipedal walking in the trees.

If this interpretation of *Danuvius* is accepted, we no longer have to come up with an explanation for why bipedalism evolved from knuckle-walking, because it didn't. It wasn't a new locomotion at all, just an old locomotion in a new setting. Put another way, the March of Progress might be backward. We didn't evolve bipedalism from a knuckle-walking ancestor. Instead, knuckle-walkers evolved from apes that were, at least occasionally, bipedal.

Another intriguing piece to this puzzle comes not from the feet and legs of our ancestors, but from their hands.

Chimpanzees have relatively short thumbs and long, hooklike fingers. Humans, on the other hand, have short fingers and a long, robust thumb. The pad of our thumb can touch the pad of each finger, giving us the often-celebrated opposable thumb. For a century, scientists have been trying to figure out how this evolved from a chimpanzeelike hand. Sergio Almécija, a paleoanthropologist at New York's American Museum of Natural History, thinks we have this backward, too.

His 2010 analysis of the 6-million-year-old thumb of *Orrorin tugenensis* revealed it to be surprisingly humanlike. Five years later, he analyzed hand bone proportions in fossil apes and humans and concluded that the human hand has changed surprisingly little in the last 6 million years. Instead, Almécija proposed, chimpanzees and the other apes evolved longer digits to keep them from falling out of the trees.

What does this have to do with bipedalism?

Walking bipedally along the tops of branches presents a balance problem. Sure, the grasping big toe helps hold on to the tree limb, but wouldn't these early bipedal apes also need powerful grasping hands? Perhaps not, say Susannah Thorpe and her colleagues at the University of Birmingham in the U.K. They found that just a light touch with the fingertips helps people balance and reduces muscle activity by 30 percent.

While the novelty of flipping the March of Progress is attractive, it's possible we are overthinking this. Can the hypothesis that we evolved from knuckle-walkers be saved? Sure. If paleoanthropologists in Eastern Africa discover a skeleton of a knuckle-walking ape in 10- to 14-million-year-old sediments, the hypothesis that the human lineage arose from a knuckle-walking ancestor would be back on the table. Those European apes, including *Danuvius* and *Rudapithecus*, might end up being extinct cousins that are leading us astray.

But science has yet to find evidence of such an animal. Still, it would be foolish to think we have this figured out. There are more fossils yet to be found and much of the human story yet to be written.

For the time being, however, the fossils we have tell a fascinating tale. From roughly 4 to 7 million years ago—more than the first third of our evolutionary history as hominins—an ape well equipped for life in the trees had followed the receding forests from Europe and scattered throughout the patchy woodlands of Central and Eastern Africa. Like its *Danuvius* predecessors, it was bipedal in the trees and could sometimes move bipedally on the ground. It had extended legs and hips like ours and could push off the ground with the outside of its foot, still retaining a grasping big toe on the inside. Different kinds evolved—*Sahelanthropus*, *Orrorin*, *Ardipithecus*, and surely many more we have not yet discovered.

For millions of years, they filled an environmental niche on the perimeter of the shrinking African forests, eating fruit and sleeping in trees. The trees were their life, but to get from one patch of trees to the next, they moved carefully and cautiously across a dangerous landscape on two legs.

But evolution does not stand still. The age of *Australopithecus* was about to begin.

PART II

Becoming Human

HOW UPRIGHT WALKING LED TO
TECHNOLOGY, LANGUAGE, THE FOOD WE EAT,
AND THE WAY WE RAISE OUR CHILDREN

Homo sapiens didn't invent bipedalism.
It was the other way around.

—ERLING KAGGE, EXPLORER,
WALKING: ONE STEP AT A TIME, 2019

CHAPTER 6

Ancient Footprints

There is charm in footing slow across a silent plain.

—*John Keats, "Lines Written in the Highlands*
After a Visit to Burns's Country," 1818

Northwest of the Ngorongoro Crater in northern Tanzania is a beautiful but stark landscape called Laetoli. The name comes from a delicate plant that produces a red-petaled flower found only in this area of Tanzania. The Laetolil flower is surrounded by plants that are not nearly as inviting.

Among the five species of acacia known in the area, two dominate the landscape. One is the classic African umbrella thorn tree seen silhouetted against a sunset in every African nature documentary. Stiff thorns protect its leaves from the probing lips of giraffes. The other acacia is a short shrub with two-inch thorns that grow out of round, black bulbs shaped like bombs in a Bugs Bunny cartoon. These hollow bulbs make a high-pitched sound when the wind blows through them, giving this acacia the nickname "whistling thorn." The bulbs also produce a sweet nectar that supports colonies of ants. Brush against them and the ants, energized with nectar, swarm out with open jaws.

It was a thorn from a whistling acacia that was embedded deep in the Maasai girl's foot.

THE MORNING DRIVE to Laetoli from our tent camp near En-dulen village took us past a herd of more than a thousand zebras. The nursing foals sported fuzzy brown and white stripes. The adults wore the familiar black and white stripes hypothesized to confuse the compound eyes of biting flies. The landscape matched the pat-tern of the zebras. The dirt road—composed of the weathered sed-iments from volcanic eruptions that have occurred sporadically for millions of years along this eastern edge of the African Great Rift Valley—was the color of charcoal. On either side of the road, wheat-colored grasses extended for miles, fed upon by zebras, wildebeests, gazelles, and cattle belonging to the pastoralist Maasai.

Seasonal rains have eroded the landscape, carving deep gullies and valleys that expose ancient rock. Charles Musiba, a stately, deep-voiced paleoanthropologist at the University of Colorado, Denver, has been finding fossils here for decades.

As a high-schooler growing up in the Lake Victoria region of Tanzania, Musiba discovered his love for paleoanthropology when he attended a talk by Mary Leakey. After hearing her stories of dis-covery at Olduvai Gorge, he approached her to ask how he could get involved. She asked if he had any skills. Well, he responded, he could draw. Leakey, who got her start as a scientific illustrator, gave the kid a chance. A few sketches of stone tools later, Musiba was a member of Leakey's team.

In June 2019, Musiba and I brought our University of Colorado and Dartmouth College students to Laetoli. Musiba's team had found an ancient hominin jawbone there a few years earlier, and we returned to the spot to see if more could be found. We had been slowly scouring the ground for an hour or so when six Maasai chil-dren appeared with a little girl no more than five years old. She was limping terribly and crying. She was swollen past her ankle, and a black crust had built up on the delicate arch of her small, bare foot. Grateful that my own young daughter had packed me a first-aid kit before I left for Tanzania, I handed our Maasai field assistant Josephat Gurtu tweezers and antiseptic wipes.

The child sobbed and buried her head into her sister's shoulder as Gurtu probed deep for the source of her pain. Soon, he extracted an acacia thorn the length of a roofing nail. An ungodly amount of pus drained from the opening. The thorn had been there for some time, and her foot was terribly infected. Gurtu washed the wound and applied antibiotic cream from the first-aid kit before wrapping it in sterile gauze and medical tape. Her sister thanked us in Maa, the Maasai language, and the young girl limped off with the other children.

I spent the rest of the morning looking for fossils but kept thinking about that little girl. The zebras and giraffes walk on keratinized hooves. Elephants have thick, padded feet. A bipedal child's soft, bare foot is vulnerable on this unforgiving landscape.

Yet, our ancestors' unshod feet have been walking at Laetoli for over 3.5 million years. The evidence of that fact, it turns out, was all around us.

IN 1976, MARY Leakey assembled a team to search for fossils at Laetoli. She and her husband, Louis, had visited forty years earlier and had found a few fossils, but their work took them to another Tanzanian site, Olduvai Gorge, which had held their attention ever since. After Louis passed away, Mary turned her attention back to Laetoli. A geological study had revealed it to be quite a bit older than Olduvai, and her team had some success finding hominin fossils eroding out of the roughly 3.5-million-year-old volcanic sediments. But even Mary could not have foreseen the wonders Laetoli was hiding in its ancient rock layers.

All it took to discover them was an elephant dung fight.

On July 24, 1976, Mary Leakey hosted a few visiting scholars—Kay Behrensmeyer, Dorothy (Doty) Dechant, Andrew Hill, and David (Jonah) Western. Leakey's son Philip showed them around the site, but eventually Western, an ecologist and conservation biologist who would later become director of the Kenyan Wildlife Service, got bored. He picked up a large piece of dried elephant dung and flung it like a Frisbee at the others. Hill, a paleontologist at the National Museums of Kenya who would go on to become an esteemed Yale University professor, returned fire. The battle continued until Hill and Behrensmeyer, now curator of paleontology at the Smithsonian Institution, took refuge in a gully. Looking for more ammunition, they spotted strangely shaped impressions in a layer of hardened volcanic ash eroding out of the hillside.

Fossil elephant footprints? Hill wondered out loud. A truce was called in the dung fight, and the others came to see what had been found.

Throughout this gully and extending beyond were footprints of antelopes, zebras, giraffes, and even birds. There were also small, odd dimples in the gray volcanic crust. Hill had seen indentations like these before in Charles Lyell's three-volume *Principles of Geology*, published between 1830 and 1833. One volume contained Lyell's illustration of fresh raindrop impressions made in mud at the Bay of Fundy in Nova Scotia, along with his argument that similar, hardened impressions found in rocks had been made by raindrops eons ago. Lyell believed that the processes affecting the Earth's surface today are the same ones that sculpted the landscape in the past. This idea, called uniformitarianism, is now a bedrock of modern science. And here, in the volcanic ash at Laetoli, Hill had found a textbook example.

For the next few weeks, the team excavated what became known as the A-site, removing the overlying sediment to reveal thousands of footprints in the hardened ash. Fossils speak broadly about the life of an organism; fossil footprints are snapshots of an individual's life.

Geologist Dick Hay got to work to figure out how this all happened. He eventually determined that a volcanic eruption had covered the landscape with a thick layer of ash. Rain turned the ash into muck. Animals walked through it for a few days, and when the sun came out, it dried and hardened like cement, preserving this moment in time 3.66 million years ago. More ash from subsequent volcanic eruptions covered the footprint layer like a blanket.

Simon Matalo, a local Maasai, could identify the footprints better than anyone. He pointed out those left by ancient elephants, rhinoceroses, zebras, antelopes, large cats, baboons, birds, and even millipedes. Most belonged to small antelopes and rabbits.

Mary Leakey told the team to keep an eye out for bipedal prints. Maybe they'd get lucky. In September, they did. Conservation biologist Peter Jones was with Philip Leakey when they discovered five consecutive prints made by something moving on two, rather than four, legs. But these bipedal prints were strange. They were small and appeared to be from something cross-stepping, meaning that the left foot stepped over the right foot like a model on a runway.

Plaster casts were made of the prints, and Mary Leakey brought them to footprint experts in London and Washington, D.C. Some of them doubted the prints were made by a human ancestor. Perhaps they were made by an extinct bipedal bear, one eventually suggested.

Mary was disappointed, but not for long.

Two years after the footprints were discovered, Paul Abell, a University of Rhode Island scholar with an expertise in geochemistry, was strolling through a neighboring area now known as the G-site when he noticed what looked like a human heel impression. Returning to camp, he reported it to Mary Leakey, but she was recovering from a broken ankle and did not want to trek out there only to be disappointed again. She sent Ndibo Mbuika, who had worked with the Leakeys at Olduvai, to inspect and report back.

Mbuika, who had discovered the very first tooth from *Homo habilis* back in 1962, didn't have to dig for long before recognizing that the footprint was from a hominin. Better yet, one print led to more. Ultimately, a total of fifty-four ancient hominin prints, proceeding along two parallel tracks, were unearthed.

These footprints are among the most remarkable discoveries in the history of our science. It appears that three, perhaps even four, individuals had been walking together, all heading north. One was on the left, a larger one on the right. A third (and maybe a fourth) walked directly on top of the largest prints.

For decades, scientists have been analyzing these footprints. Most have found them to be consistent with what we know from the bones of *Australopithecus afarensis* (Lucy's species). They walked with a prominent heel strike and with the big toe in line with the other digits. There also appeared to be an incipient arch in the foot. In other words, *Australopithecus* had a foot a lot like ours and walked like we do, with a few subtle differences.

When humans walk today, we transfer weight onto our big toe and push off the ground primarily from there. The Laetoli prints were made by individuals who also pushed off from the big toe but kept some weight on the outside of the foot as they did so. The footprints show that the Laetoli hominins would have been flat-footed by modern human standards, something also inferred from the bones this species left us. It appears, too, that the Laetoli footprints were made by individuals who didn't drive off the ground with as much force as humans do today, although that could also be because they were walking through thick, wet ash.

Unlike expectations that one might have from the March of Progress image, *Australopithecus* did not walk crouched over like a chimpanzee. They didn't wobble back and forth like Louis the gorilla when he carries his tomatoes from one side of his enclosure to the other. Instead, they walked tall, with an extended leg and an extended hip. Like us.

The Laetoli footprints capture a moment in the life of our extinct relatives. The footprints are close together and in sync, revealing that the individuals were walking slowly in a group. They were heading north, toward the Olduvai Basin, where there would have been water and tree cover. Because foot size and height are roughly correlated, we can estimate that the smaller individual to the left was just under four feet tall, about the size of Lucy. The larger one on the right was just under five feet. The smaller individual's right foot landed at an odd angle, suggesting that this hominin may have been injured. Toward the end of the trail, the prints are jumbled and slightly deeper as the group pivoted and turned. After a pause, they continued north.

The Laetoli footprints are an elegant example of the intersection of science and imagination. Some things we know for sure—that they come from bipeds, for example. But were they holding hands? Did they have their arms around one another as depicted in the diorama of Laetoli on display at the American Museum of Natural History in New York? The questions are mesmerizing and haunting. Imagine:

We must go north. That is where the water is and where the others should be. Yesterday was hell. The air was thick. The sky turned black and rained ash. The ground shook. The earth growled like a predator. Today is better, but I am scared and hungry. The clouds are still thick, and they still flash and rumble, but at least water, instead of ash, is falling. There is nothing to eat. The grass is gone, covered by ash. We have to go north. To the water. There will be food there and the others, unless they, too, are covered by the ash. It clings to the branches of our trees. We carefully scan the landscape and climb down, stepping onto the slushy surface. We walk single file, stepping in each other's footprints, masking our numbers. The ground is slippery, and we have to walk slowly. Plus, Mum is still hurt. She walks to our left, gingerly. A thorn is buried deep in her foot. We cannot get it out. Zebra footprints cross our trail. We see a group of guinea fowl and some dik-diks searching for food. An elephant eyes us closely. As the ground rumbles again, Mum stops and pivots to the west. Nothing. We carry on to the north, slowly plodding through the thick ash. North, water.

THE FOOTPRINTS AT Laetoli can be a bit confusing, so let me explain. The famous fifty-four-footprint parallel trackway I have been describing was found at G-site. But recall that the very first bipedal footprints discovered at Laetoli were strange. They were found at A-site and looked like they were made by a runway model. Some researchers thought they may have been made by an extinct species of bear. I was eager to solve the mystery of these odd bipedal footprints. But first, we had to find them.

I FELT FAINT. We had been excavating all morning at the A-site. Now the hard sun was directly overhead. I hadn't had enough to drink, my Nalgene half full of warm water. Just below my wedding ring, a blister had formed, grown, and burst after hours spent scraping the hardened topsoil from the footprint layer. Using Mary Leakey's old maps, I had measured to the precise location where the mysterious bipedal prints should be. Blaine Maley, an anatomy professor at the Idaho College of Osteopathic Medicine, and Luke Fannin, a Dartmouth graduate student, had found and carefully re-excavated the cluster of baby elephant prints from Mary's map. The A-prints should have been just four meters to the west, but there was nothing there except eroded dik-dik and guinea fowl footprints. Had erosion, which initially uncovered the bipedal prints, washed them away sometime in the last forty years?

I joined a few of my students and Shirley Rubin, a professor at Napa Valley College, under an umbrella acacia, our only source of shade. A ball of angry bees hung from a nearby tree, filling the air with an unsettling buzz. Should we give up on the A-prints? Could they have survived forty years of seasonal rains?

A graduate student in my Dartmouth College paleoanthropology laboratory, Ellie McNutt, whom we met earlier, had recently defended her Ph.D. thesis and begun teaching anatomy at the University of Southern California medical school in Los Angeles. An Iowa native, she approached science with midwestern sensibility and was skeptical of the accepted wisdom that an ancient bear had made the Laetoli A-prints. No fossil bears had ever been found at Laetoli, and the A-prints were so small that they would have to have been made by a cub walking like it was in the circus. Studying decades of film taken by American black bear expert Ben Kilham at his Lyme, New Hampshire, study site, she learned that when people aren't around, bears don't walk bipedally very often. Only once in fifty hours of video footage did a single bear take five consecutive upright steps—what would be required to replicate the A-prints.

To further investigate, McNutt had dangled syringes of maple syrup above the heads of juvenile black bears being rehabilitated by Kilham for release into the wild, coaxing them to stand and walk bipedally through wet mud. She measured the prints they made and compared them to photographs and published measurements of the A-trail. It was no match. Bears cannot balance at the hip like we can. They wobble from side to side, producing widely spaced footprints. But the A-prints were narrow, even cross-stepping. As Mary Leakey had admitted, the A-prints were never fully excavated. Unless we could find them again, claiming that they were made by hominins would be a tough sell.

I chugged the rest of my water, plunked my sun hat on, and walked back to the excavation, where our field assistant Kallisti Fabian was still scraping the hard soil with the flat edge of his trowel.

"Mtu," he said.

"What?"

"Mtu!"

Fabian had finally reached the ash layer. There, he had uncovered a small depression. It was a footprint.

"Mtu?" I asked, repeating the Swahili word for "human."

"Ndiyo," Fabian replied in the affirmative.

I dropped to my belly, pulled out dental tools, and began to gently pull the overlying matrix from the print. Like a cookie crust, the layer peeled away from the ancient volcanic ash, revealing a small, beautifully preserved hominin footprint. I touched the impression made by the heel and gently rubbed my thumb along the rim of the big toe.

This was no bear.

I have devoted much of my career to studying the evolution of upright walking but had never put my hands or eyes on an actual *Australopithecus* footprint.

Until then. A chill ran down my spine.

"That's it! That's the A-trail!"

I was thrilled. Dartmouth College graduate student Kate Miller and undergrad Anjali Prabhat came over and helped continue the excavation, searching for additional prints. I had all but given up, fearing the original A-prints had washed away, but here they were. Over forty years of erosion had not destroyed the footprints; it had covered and preserved them.

Josephat Gurtu walked over and smiled.

"Mtu?" I asked.

He nodded. "Mtoto," he said. That means "child." That could be why the A-prints looked unusual, why they were small, and why they were cross-stepping. They may have been made by a hominin child.

The foot that made the A-prints was a little over six inches long, about the size of the foot of a four-year-old modern human child. The big toe made a clear impression in the hardened volcanic ash, and there was a prominent heel strike.

That afternoon, we returned to the site and excavated the rest of the trail, rediscovering five consecutive footprints. We 3D-laser-scanned the prints so that there would be a digital copy for study and for posterity. We extended the excavation back and to the side but did not find any more. The wet volcanic ash layer must have been hardening when this hominin walked across the landscape like a ghost, leaving no further trace. But here, at this one spot, the ash must have still been wet (perhaps it was in the shade of a tree), and five prints were recorded.

While the students cleaned the prints, they played 1980s music from their phones—Bruce Springsteen, Yes, Don Henley, Hall & Oates. By the late afternoon, the Maasai children were back. The little girl with the infected foot looked much healthier and happier, although she was still wary of us and hid behind her sisters. She had visited the hospital in Endulen for proper treatment, her foot now covered by bandages and shod with a Maasai tire sandal. Her foot was about the same size as the one that made the A-prints. A small hominin had walked in this very spot 3.66 million years ago.

I opened the compass app on my phone and angled it in the direction of the trail. After the ash fell from the sky, the young hominin had trudged north just like the others.

Bipedal walking in the earliest hominins—*Sahelanthropus, Orrorin*, and *Ardipithecus*—is controversial. These early members of our lineage, living between 4 million and 7 million years ago, were still adapted for life in the trees. Although the fossil evidence contains tantalizing hints of bipedalism, the degree to which they walked that way on the ground is debatable until we discover more fossils and new methods for studying these ancient bones.

Lucy established the antiquity of upright walking at 3.18 million years, but interpreting her old bones has not been easy either. The Laetoli footprints, however, leave no doubt. Early members of the genus *Australopithecus*, living more than 3.6 million years ago, walked on two legs much like we do today. Fossil evidence also indicates that *Australopithecus* was a capable tree climber, which makes sense. Without a campfire to gather around or structures to sleep in, there was only one place to stay safe at night: the trees. During the day, however, *Australopithecus* was on the ground, walking on two legs, foraging for food, just trying to survive.

But bipedalism is about more than getting from one place to another. It served as a gateway for other changes—critical ones that help make us who we are today.

BEFORE DISCOVERING THE footprints at Laetoli, Mary and Louis Leakey spent decades at Olduvai Gorge, collecting hundreds of stone tools they grouped together in a culture called the "Old-owan." Radiometric dating of Olduvai revealed the tools there to be about 1.8 million years old. In 1964, the Leakeys found their elusive toolmaking species. It was a hominin with a humanlike hand and a slightly larger brain and smaller molar teeth than an *Australopithecus*. They designated the fossils they found as a new species, *Homo habilis*, which roughly translates to "handy man."

Just a decade later, Don Johanson discovered Lucy, and a few years after that, Mary Leakey's team found the Laetoli footprints, pushing bipedalism back to at least 3.6 million years ago. Bipedal walking, it seemed, was twice the age of the oldest stone tools.

Darwin, it was argued, was therefore wrong. In *The Descent of Man*, he had hypothesized that bipedalism and stone toolmaking evolved in concert, uprightness freeing the hands to create the tools that rendered large canines unnecessary and jump-starting the growth of the brain. But based on Lucy and the Laetoli footprints, upright walking long preceded stone tool development.

However, more recent discoveries have invited us to revive Darwin's ideas from the dustbin of science.

In 2011, Sonia Harmand, an associate professor of anthropology at Stony Brook University in New York, was conducting an archaeological survey along the west side of Lake Turkana in Kenya, heading to an area called Lomekwi near deposits where a team led by Meave Leakey, another member of paleontology's first family, had discovered a 3.5-million-year-old hominin skull in 1999 that she'd named a new species, *Kenyanthropus platyops*. Harmand and her team took a wrong turn, leading them to new outcrops. Harmand got out of their vehicle and invited the team to survey this new area. After one hour of prospection, the team spotted some odd stone tools eroding out of the sediments.

They mapped the stone tools found on the surface, and the following year Harmand's team discovered 150 stone tools that had been created by hominin hands. The rocks used to make the tools were bigger and a lot simpler than those found at Olduvai Gorge. It looked like the hominins who made these tools at Lomekwi had just picked up some large rocks, smashed them together using primitive percussion techniques, and grabbed the sharp pieces that flaked off. Simpler sometimes means older. Sure enough, these tools were sandwiched between volcanic tuffs deposited 3.3 million years ago.

After the Leakeys' work at Olduvai, older Oldowan tools that date back to 2.6 million years were discovered in Ethiopia. Harmand's discovery pushed stone tool technology back an additional three-quarters of a million years, out of the hands of *Homo habilis* and securely into the hands of *Australopithecus*.

Meanwhile, on the other side of the Awash River from where Lucy was discovered, Zeray Alemseged of the University of Chicago was working in an area called Dikika. There, he discovered an extraordinary partial skeleton of an *Australopithecus* toddler—what the media called "Lucy's baby." In 2009, Alemseged invited me to study the foot of this rare find, and we published our results in 2018. The foot was a lot like a human's, indicating that the Dikika Child was already walking on two legs. But it also preserved evidence for a more mobile big toe that would have allowed her to scramble into the trees or hold on to her mum more effectively than a human child can today. It made sense. Drive by any elementary school playground and you are likely to see the kids climbing, just as they did millions of years ago.

In these 3.4-million-year-old sediments, Alemseged also unearthed a few fossil antelope bones that had been deliberately cut by sharp rocks. Our small *Australopithecus* ancestors were not capable of hunting anything that large, but these cut marks indicate that they were able to use sharp rocks, like the ones Sonia Harmand found at Lomekwi, to scavenge some meat from a kill.

Tools are a game changer. What started as a way for *Australopithecus* to dig for roots and tubers and to hack meat from the remains of lion kills over 3 million years ago has culminated in iPhones, antibiotics, ballistic missiles, and the New Horizons space probe. In one of the greatest technological achievements in the history of our species, we celebrated upright walking on another world as "one small step for [a] man; one giant leap for mankind."

Every human culture uses tools. Our bodies have become biologically adapted to a diet and a way of life possible only with technology. That shift began with *Australopithecus*, the first habitual biped and the first user of stone tool technology. Bipedalism freed the hands from the requirements of locomotion. Freed hands could smack rocks together to create a cutting edge. That cutting edge allowed a hominin to acquire previously inaccessible food. The increased energy a better diet provided eventually launched our lineage to the very edge of the solar system.

To be sure, chimpanzees also make tools. Jane Goodall's famous report of chimpanzees peeling tall grass to create tools for extracting termites from their mounds had forced scientists to reconsider the long-held belief that toolmaking was exclusively human. As Louis Leakey famously wrote: "I feel that scientists holding to this definition are faced with three choices: They must accept chimpanzees as man, by definition; they must redefine man; or they must redefine tools."

Since that time, monkeys, crows, otters, puffins, some species of fish, and even octopuses have been observed using tools. But no other species relies on them the way that we do, and this reliance began soon after we had liberated our hands by walking on two legs.

How closely terrestrial bipedalism and the invention of stone tools are connected in time remains uncertain. The footprints at Laetoli are about a quarter of a million years older than the oldest reported evidence for stone tool use in *Australopithecus*. The 4.2-million-year-old shinbone from Kanapoi, Kenya, attributed to the first *Australopithecus,* called *anamensis*, is very humanlike, widening the gap between walking and stone tools to at least 800,000 years. But it is possible that *Australopithecus* at that time was using its freed hands to make wooden digging sticks or to fashion baby slings from vines and palm leaves. Such plant material would not have survived in the archaeological record.

Darwin may not have been entirely right that walking and stone tool use happened in lockstep, but he was probably closer than we once thought.

WITH OUR FREED hands, though, there was more to carry than just tools. The evolution of upright walking, it turns out, fundamentally altered how we parent our children.

In 2010, my wife gave birth to twins. During those first few exhilarating and exhausting weeks, I was struck by two things (other than the realization that I had married a superhero): how desperate we were for help and how willing people were to provide it. It made me wonder what Lucy would have done with a newborn baby in a much more hostile environment.

While holding my helpless newborns, I would sometimes imagine that I was Lucy. I pictured myself as a female *Australopithecus* living over 3 million years ago. My environment did not consist of houses, Dunkin' coffees, roads, and squirrels. Instead, there were vast expanses of open grassland, rivers, and predators—big ones like *Homotherium*, a saber-toothed cat larger than a modern lion. As Lucy, I walked on two legs, making me one of the slowest animals in the landscape. I spent much of my day foraging on the ground for termites, fruit, roots, and tubers. I was almost always hungry. I could climb trees reasonably well, but not nearly as well as a chimpanzee. My arms were long enough to reach my knees, but not as long and powerful as an ape's. Life as Lucy was hard. It always had been. But now it was a lot harder with a baby in my arms.

A chimpanzee's baby rides on its mother's back when she knuckle-walks through the forest. It holds on to her fur with strong hands and a grasping big toe when she climbs. Lucy's baby would not have been able to do that. As an *Australopithecus*, Lucy stood upright. If she put a baby on her back, it would have slid right off. And an *Australopithecus* infant's toes weren't strong enough to hold on when its mother climbed.

Evidence from the genetics of the three different lice species that reside on humans (head lice, pubic lice, and clothing lice) suggests that our hominin ancestors may have begun to lose their body hair around Lucy's time, so there may not have been much for a kid to grab on to anyway. Lucy would have had to carry her helpless infant in her arms.

Every morning, she would have climbed down from her tree to search the grassland for enough food to keep her and her infant alive. When her baby was fidgety, she would have nursed it to keep it quiet. The slightest cry could have perked the ears of a nearby *Homotherium*. Every member of her group would have been vulnerable, constantly scanning the horizon for predators. In the evening, as nocturnal predators emerged from their slumber, she would have climbed back into a tree for the night. But it would have been difficult, perhaps impossible, to climb with one arm while cradling her baby in the other. What could she do? By evolving bipedalism, our ancestors faced new challenges—challenges that required new solutions.

We'll come back to Lucy in a moment. For now, consider the first few months in the life of a chimpanzee. A chimpanzee typically gives birth at night and in solitude. The baby can hold on to its mother's fur soon after birth, and the mother carries it everywhere for the next six months or so, rarely letting any other member of her group touch it.

The first six months of a human's life are quite different. Birth is typically a social event, with female helpers, or midwives, assisting. The newborn is helpless, and the mother benefits from others who help feed it, coddle it, smile at it, speak to it, and keep it safe. As the saying goes, it takes a village.

How did that happen? How did we humans become a species that recruits helpers to assist with the raising of our young?

We left Lucy at the base of a tree, wondering how she could climb it with one arm while holding on to her helpless infant with the other. The most obvious solution is that she handed it off to another in her group. Perhaps she also did this while she foraged. Maybe an older child held the baby. Perhaps it was a sister—the baby's auntie. It could even have been an unrelated female—a friend. Maybe the father held the child while the mother foraged and slept. These small acts of parenting by others require trust, cooperation, and reciprocation.

These qualities are fundamental to society today, but they have their roots in solutions to the problems faced by *Australopithecus* because they moved on two, rather than four, legs. We can trace the way we collectively raise our children today back to Lucy and her kind.

SAFARI OUTFITS WAKE their clients before sunrise to take them on morning game drives. They head into the game parks again in the late afternoon and even offer night drives, using spotlights to catch the glowing eyes of Africa's diverse wildlife. A noon game drive would fail to reveal any wildlife. The only animals out and about then are people. This, it turns out, is telling.

The shift to bipedalism made us slow and vulnerable to predators, yet the fossil record of Africa at this time teems with huge things that wanted to eat us. Fossils of two different enormous saber-toothed cats (*Homotherium* and *Megantereon*) and another as big as a large leopard (*Dinofelis*) have been found in places where *Australopithecus* has been discovered. Paleontologists have also collected fossils of a hyena roughly the size of a modern lion (*Pachycrocuta*) with powerful, bone-crushing jaws and teeth. Large crocodile fossils are also common. From the tooth impressions found on the occasional hominin fossil, we know that these large predators posed a constant threat to our ancestors, but only occasionally were they successful in nabbing one.

Faced with similar threats today, primates cluster in large groups. When a leopard stalks a group of baboons, the chances are that one of the baboons will spot the predator and sound the alarm. This works only if there are enough vigilant baboons. Two or three baboons focused on finding breakfast may miss the enormous ghost cat slinking toward them and become breakfast themselves. Living in a large group also minimizes the chance of any particular baboon being picked off by a successful predator. The odds are 50 percent in a group of two, but only 2 percent in a group of fifty. But baboons can run fast and climb quickly. *Australopithecus* could not. How did we avoid becoming a regular meal for these large cats and hyenas?

When the sun is overhead and the temperature rises, modern lions, leopards, cheetahs, and hyenas find some shade and go to sleep. Even prey—antelopes and zebras—lie in the tall grass to stay cool. Our ancestors' option for survival was to avoid being on the ground when predators were active—dusk, night, and dawn.

Australopithecus must have been active during the day, just like I am. After the cats and hyenas had filled their bellies and found shady spots to rest and digest, *Australopithecus* would have climbed down from their trees to search for food. They would have eaten whatever they could find—fermented fruits, nuts, seeds, tubers and roots, insects, young leaves, and occasionally scraps of meat hanging off discarded overnight *Dinofelis* kills.

Today, humans around the world collectively eat everything. If it has DNA, we have tried it. This shift toward a generalized diet appears to have started early in our evolutionary history. Walking on the ground made us vulnerable. We couldn't afford to be picky eaters.

Australopithecus moved in large groups, eyes constantly scanning for any subtle movement in the tall grass. If they encountered a predator, fleeing at twenty miles per hour would have been fatal. *Homotherium* would have had plenty of time to chuckle and lick its chops before running down and eating one of them. Today, baboons and chimpanzees, when left with no other options, mob a would-be predator, standing tall, shrieking, and sometimes hurling rocks or sticks to drive it off. It seems all but certain that *Australopithecus* did the same, cooperating with one another against a common enemy.

Moving around equatorial Africa at high noon presents other challenges. For one thing, it is hot. If our ancestors stayed in the woods, this would not have been as big of a problem. But there is evidence from the carbon isotopes in the teeth of *Australopithecus* that a large percentage of their meals came from grasslands. How did they stay cool?

Remember, our body hair became patchier and finer around the time of *Australopithecus*. Perhaps concentrations of sweat glands also increased as the skin of *Australopithecus* was exposed to the air and could cool the body, but sweat gland evolution remains poorly understood. After a day foraging together, our *Australopithecus* ancestors would have made their way back to the trees to groom, cuddle, and sleep.

In 1871, Darwin proposed that the increasing size of our ancestors' brains was the consequence of a suite of changes in early members of our lineage—bipedalism, tool use, and canine reduction. But the timing doesn't appear to work. The earliest bipeds had brains that were no larger than a modern chimpanzee's. Other researchers have proposed that walking on two legs required a large brain to balance and coordinate such a sophisticated musculoskeletal machine. Tell that to a chicken, whose brain is the size of an almond.

Clearly, bipedalism and brain size do not go hand in hand; yet, in us, they kind of do. How did that happen?

The brains of the earliest hominins, *Sahelanthropus* and *Ardipithecus*, were about the size of the average chimpanzee brain—375 cubic centimeters (cc), or slightly larger than the volume of a can of soda. Recently, paleoanthropologist Yohannes Haile-Selassie discovered a magnificent skull of a 3.8-million-year-old *Australopithecus anamensis*. It, too, had a similar-size brain. But by the time of Lucy, half a million years later, brains had increased to an average of 450cc. That's still only a third the brain size of an average modern human, but the jump from 375cc to 450cc was a 20 percent increase in volume, and brains don't come cheaply.

Your brain is only 2 percent of your body weight, but it uses 20 percent of the energy you take in. That means every fifth breath you draw and every fifth bite of food you eat is earmarked for the hungry cells of the brain. How, then, could *Australopithecus* afford this increase in brain size?

Treadmill studies demonstrate that chimpanzees use twice as

much energy as humans when they move. Apes also use a lot of energy when they climb, which they do frequently. Because they need so much energy just to move around, they don't have excess energy to devote to an inflated brain.

Perhaps that changed with *Australopithecus*.

Perhaps, as our ancestors moved on two legs and climbed trees less often, they had an energy surplus. Perhaps those among them who had slightly larger brains put that excess energy to good use, navigating complex social situations in their group more effectively. Maybe they solved foraging problems with newly invented stone tool technology. This, in turn, might have increased their ability to find food, making still more energy available for brain growth.

A variant on this idea is not that there was a surplus of energy, but that the efficiency of bipedal walking allowed our ancestors to have a wider search radius as they foraged for food. In a grassland environment, food may have been more spaced out and could only have been accessed by a hominin that moved with more efficiency than a standard ape.

Those are all evolutionary explanations for *why* brains got larger. But we can also ask a more basic question. *How* did *Australopithecus* grow larger brains than their predecessors?

Humans have bigger brains than chimpanzees for two reasons. First, we have a faster rate of brain growth; we simply add more brain tissue each year as we mature. Second, we grow our brains for longer, reaching full brain volume by age seven or eight compared to three or four in chimpanzees.

What happened in *Australopithecus*? The skull of a young child provides the answer.

Discovered in 2000 by Zeray Alemseged, the Dikika Child has a well-preserved skull and a sandstone impression of the brain. High-resolution scans of the skull were taken at a huge particle accelerator in Grenoble, France, which produces X-rays with enough power to penetrate the fossilized skull and see extraordinary details of the brain and unerupted teeth of this child.

Tanya Smith, a human evolutionary biologist at Griffith University in Australia, used these high-resolution scans to measure the incremental growth of the Dikika Child's teeth as if they were tree rings. Her clever analysis revealed that the child died when she was two years and five months old. At this age, the child had grown a 275cc brain, which is about 70 percent the size of adult *Australopithecus afarensis* brains. In contrast, chimpanzees of the same age have close to 90 percent of their brain growth complete. Humans grow their brains slowly, and the Dikika Child's fossil indicates that Lucy's species did as well. As happens in humans, *Australopithecus* youngsters' growing brains formed connections as they learned what to eat, what threats to avoid, the relationships between group members, and other skills critical to their survival.

In animals under intense predation pressure, natural selection favors *rapid* development. The idea is to grow and reproduce quickly while you are still alive. Only animals that are rarely predated upon—elephants, whales, and modern humans—can afford to have slowed-down growth and long childhoods. Slow brain growth in *Australopithecus* indicates that our hominin ancestors developed a buffer—likely a social one—against predation. We looked out for one another.

Sure, we occasionally lost one of our own to a *Pachycrocuta*, but in general, our ancestors must have been buffered enough that slowed brain growth, and the resulting emphasis on learning in young *Australopithecus* individuals, was favored by natural selection.

However, this did not turn into a feedback loop—not yet. The brains of *Australopithecus* plateaued at 450cc and did not budge for more than a million years. Bipedalism did not evolve in sync with stone tool technology or with an increase in brain size, as Darwin envisioned. But it opened the door to these new possibilities.

New discoveries have complicated this story, however. Now when someone asks me about bipedalism in *Australopithecus*, I ask, "Which *Australopithecus*?"

CHAPTER 7

Many Ways to Walk a Mile

There is more than one way to skin the bipedal cat.

—*Bruce Latimer, paleoanthropologist, Case Western Reserve University, 2011*

In the summer of 2009, I was working in the fossil vault in the anatomy department at the University of the Witwatersrand in Johannesburg, South Africa. My flight back to the United States was departing in a few hours, and I still had several more *Australopithecus* foot bones to 3D-laser-scan.

Suddenly, South African paleoanthropologist Lee Berger burst into the room, a crazed look on his face.

I had never met him, but I knew of him and his work. He was, after all, a prominent figure in my field. Clearly, he hadn't come looking for me or the other young researcher in the room, Zach Cofran, now a professor at Vassar College in New York. Berger scanned the room, looking past Cofran, me, and the dozens of hominin fossils laid out on the table. Then we locked eyes.

"Do you want to see something cool?" he asked.

Cofran's eyes lit up at the invitation, but I froze. Panicked that I didn't have time to finish my work, I sure as hell didn't have time to waste on "something cool." Then again, I'd just started a new scan, positioning a fossil on a rotating turntable while my cereal-box-size scanner slowly converted a 2-million-year-old foot bone into a digital replica on my computer screen. Once a scan was running, I *could* walk away from it for a few minutes.

"Do you?" Berger said, this time with a mischievous grin. He was so excited about sharing something that even two young researchers he didn't know would do.

"Sure," Cofran and I both said.

Berger, who had grown up in Georgia but had lived in South Africa for the last thirty years, led us down a hallway and into a lab where there was a large table covered by a black velvet cloth. What was under that black cloth would change everything I thought I knew about the evolution of upright walking.

IN OUR SCIENCE, there are field-workers who find fossils and lab scientists who analyze them. Most of us end up doing both, but rarely is there an even balance, so we develop reputations as someone who either works with a trowel or with a computer. Berger is a field-worker, an explorer—the closest thing we have in paleoanthropology to Indiana Jones. But in the early 2000s, with funding shifting from field work to new digital methods for analyzing old fossils, the discipline was transitioning.

Some even wondered if all of the important human fossils had already been found. Berger didn't think so. For close to two decades, he had been excavating at a cave called Gladysvale—one of the most fossil-rich places in the Cradle of Humankind in South Africa. Bones of ancient zebras, antelopes, warthogs, elephants, gazelles, giraffes, and baboons pour out of the walls of the cave. Berger and his small team had collected thousands of them there, but the only hint of hominins were two *Australopithecus* teeth.

Gladysvale's rich concentration of fossils was as seductive as a Siren, but for a paleoanthropologist like Berger, it could be a career killer if nothing but more antelope and zebra bones turned up. Eventually, Berger thought, he'd find his Lucy. While that turned out to be true, and then some, it wouldn't happen at Gladysvale.

Ever the explorer, Berger spent the first few years of the twenty-first century expanding his search beyond Gladysvale, surveying the area around Johannesburg for caves by using expensive high-resolution images taken by the United States military. But in 2008, this got much easier. Lee Berger downloaded Google Earth.

He spent weeks staring at his computer screen, studying satellite images of the same dry landscape he had explored for twenty years and identifying patterns he couldn't see from the ground. He saw wild olive and white stinkwood trees growing in clusters. Where, he wondered, were these water-dependent species finding water? He and geologist Paul Dirks solved the mystery.

Rainwater pools at the bottom of vertical cave shafts, where seeds from olive and stinkwood trees blow in, germinate, and take root. Reaching for the sun, they rise to the surface, revealing where the caves are. Berger recorded the GPS coordinates of the tree clusters and spent months driving around the Cradle, confirming his suspicion. Caves were everywhere, upward of six hundred that no one had previously documented. But did they contain fossils?

In August 2008, Berger explored one with his nine-year-old son Matthew, their Rhodesian Ridgeback dog Tau, and postdoctoral researcher Job Kibii, currently the curator of paleontology at the National Museums of Kenya. The cave is now called Malapa, which means "home" in the local Sotho language.

"Go find some fossils, Matthew," Berger encouraged his son. A few minutes later, Matthew was following his dog when he tripped over a large chunk of rock, picked it up, and announced, "Dad, I found a fossil."

Fossils are not uncommon in this region, but like the thousands

Berger had collected at Gladysvale, they are almost always of antelopes, zebras, or warthogs. As Berger approached, however, what
he saw sticking out of that rock was not from an antelope or zebra.
It was the collarbone of an early hominin. He turned the rock over
and saw part of a lower jaw with a small, blunt canine tooth that
was distinctly hominin. In under ten minutes at Malapa, Berger's
nine-year-old son had discovered as many hominin fossils as Lee
had found in two decades at Gladysvale.

These bones alone would have been a significant contribution to
our science, but Malapa had a lot more to offer. Over the next several months, Berger's team kept finding hominin fossils. The bones,
trapped in hard, concretelike blocks of reddish, fossilized gravel
called breccia, were transported to the laboratory at the University of
the Witwatersrand. There, trained technicians spent months with
tools that look like mini-jackhammers, slowly freeing the fossils
from the surrounding rock grain by grain. When they were done,
they had two partial skeletons.

Two ancient hominins! That was what Lee Berger had revealed,
like a magician, under the black velvet cloth.

One skeleton belonged to a young male. His bones had open
growth plates, indicating that he was about eight years old and still
growing at the time of his death. The skull was immaculately preserved. His wisdom teeth were still tucked away, unerupted in
their crypts. Called Malapa Hominin 1 (MH1), he was nicknamed
Karabo, which means "the answer," by local schoolchildren.

The other skeleton, Malapa Hominin 2 (MH2), belonged to an
osteological female whose worn wisdom teeth indicated that she
was an adult. The skeletons had several jagged fractures in their
arms, shoulders, jaws, and skulls, which occurred when the bone was
still alive. Such clues allow us to determine the probable cause of
death: they had plummeted over fifty feet into the cave shaft, died

on impact, and decomposed away from the bone-crunching jaws of scavengers.

The skeletons were sandwiched between layers of uranium-rich limestone, which can be dated because uranium is radioactive and decays to lead and thorium at a known rate. University of Cape Town geologist Robyn Pickering determined the age of MH1 and MH2 with unusually remarkable precision. She found that these hominins died within a 3,000-year window 1.977 million years ago.

Six months after my sneak peek at these fossils, Berger and a half dozen coauthors announced to the world that they were from a new species of *Australopithecus*. They named it *sediba* and proposed that it might even be the long-sought direct *Australopithecus* ancestor to our own genus, *Homo*.

Days later, Berger and foot expert Bernhard Zipfel invited me to work with them to study the foot and leg bones of this new hominin. If I had declined Berger's invitation to "see something cool," the offer probably would have gone to someone else.

Two years earlier, I had completed my Ph.D. thesis on the foot and ankle of fossil apes and hominins. I knew the foot and leg of *Australopithecus* well and was thrilled to be invited to study this new species. Every recent Ph.D. in paleoanthropology dreams of an offer like this, and what I'd seen under that black velvet cloth had intrigued me.

THE PACKAGE FROM South Africa arrived at my Boston University office just as I was racing out the door to try to beat the traffic. I put it under my arm, jumped into my little red Toyota Matrix, sped down Commonwealth Avenue, and hit a wall of traffic as I turned onto the Mass Pike. With thousands of cars inching west at five miles per hour, I could safely take my hands off the wheel and open the box from the University of the Witwatersrand.

Berger and Zipfel had sent plastic casts—exact copies—of the foot bones of *Australopithecus sediba*. With my wife six months pregnant, I wouldn't be able to get to Johannesburg again to study the original fossils until the following year, but these casts would give me a good idea of what we could say about the foot of this 2-million-year-old hominin.

I fumbled through the box, pulled out little bundles of bubble wrap, and ripped one open. Inside was a small talus, the top bone of the foot that forms the ankle joint with the shinbone. I studied it with one eye, the other on the brake lights of the car in front of me. At first glance, it was humanlike, similar to Lucy's talus in some ways, and different in others, but I didn't conclude much from that. The talus is a notoriously variable bone. The talus bones of all the Bostonians stuck on the Mass Pike with me were as different from one another as Lucy's was from *sediba*.

At the Weston junction, only two tollbooths were open where the Mass Pike merges with I-95, one of the most congested highways on the eastern seaboard. I knew I'd be stuck there for a long time, so I tore open another bundle and picked up a small end of a shinbone, turning it over and over. It resembled both a human's and Lucy's except for a chunk of bone called the medial malleolus—the rounded bulge on the inside of the ankle. The bulge was enormous, way bigger than either a human's or Lucy's. Only apes have a medial malleolus that big.

Something wasn't right. Perhaps it was a casting mistake, or even a pathology, the result of disease or injury. The car behind me honked, and I crept closer to the tolls.

I tore open another bundle and stared in amazement at a complete calcaneus, the heel bone that whacks the ground first when we walk. In humans, it is the largest foot bone—about the size of a small potato. In Lucy's species, *Australopithecus afarensis*, the heels were also large and chunky, well adapted for the forces absorbed when walking on two legs. Even the Laetoli footprints show that whoever made them had a big heel. But the funny little heel I held in my hands resembled a chimpanzee's. It didn't look like it could have come from something that walked on two legs.

Had Berger and Zipfel sent me a chimpanzee calcaneus? Was this a joke? Some kind of test to see if I was up to the job? I stared at the heel bone again and turned it in my hands, noticing the chimpanzee-like heel melded with humanlike features farther down the bone.

I had never seen a heel like this one. I was mystified and enthralled. The guy behind me leaned on his horn. Traffic sped up on the other side of the tolls, so I couldn't look at the bones again until I was home. It took the rest of that day and then another three years, including several trips to South Africa, to figure out what was going on.

THE OLDEST *AUSTRALOPITHECUS* fossils, from lakeshore sediments in Kenya and woodland soils in Ethiopia, are 4.2 million years old. The youngest, found in caves in South Africa, are about 1 million years old. In those 3 million years, *Australopithecus* diversified into many different kinds. In fact, scientists have named over a dozen.

The original *Australopithecus*, named by Raymond Dart for his Taung child, is *africanus*. Lucy's species is *afarensis*. The oldest known *Australopithecus* is *anamensis*. There are controversial species known from just a few fossils, such as *platyops*, *garhi*, and *bahrelghazali*. There are large-toothed *Australopithecus* that are referred to as the robusts. They are sometimes given their own genus name—*Paranthropus*—and come in three different forms: *aethiopicus*, *robustus*, and *boisei*.

Berger had just announced the discovery of a new *Australopithecus*: *sediba*.

Even though there were differences between the many *Australopithecus* species, one characteristic united them: they all walked on two legs. It turns out, though, to be much more complicated than that.

In the early 1970s, J. T. Robinson, who had trained under Robert Broom (the guy who used dynamite to dig for fossils), proposed that the two different kinds of *Australopithecus* found in South African caves—a large-toothed one called *robustus* and a smaller-toothed one called *africanus*—walked differently. Noting differences in their pelvises and hip joints, he proposed a shuffling gait for *robustus* and a more humanlike stride in *africanus*. But the pelvis is such a thin bone that it is easily damaged and distorted during the fossilization process. It was hard to tell whether the skeletal differences Robinson had detected were present when these hominins were alive.

Thirty years later, American Museum of Natural History paleoanthropologist Will Harcourt-Smith studied the foot bones of *africanus* and Lucy's species (*Australopithecus afarensis*) and proposed something similar. Using a method called geometric morphometrics, which could capture and quantify the complex 3D shape of foot bones, he argued that Lucy and her kind had humanlike ankle joints, but that other parts of their feet were more apelike. And that South African *Australopithecus africanus* had the opposite—apelike ankles and humanlike feet.

I was skeptical. After all, we had only a small handful of bones from each species. Were the differences Harcourt-Smith had detected any more significant than the normal variations in the foot bones of modern humans? The way I saw it, *Australopithecus* had evolved humanlike walking abilities, and any variations between species or between fossil sites in Africa were biologically insignificant noise.

I was wrong, but it took *sediba* to change my mind.

Despite being a million years *younger* than Lucy and her kind, the foot bones from *Australopithecus sediba* were, for the most part, *less* humanlike. The knee, pelvis, and lower back made clear that *Australopithecus sediba* moved bipedally, but not like we do and certainly not like other species of *Australopithecus*.

In 2011, after an exhaustive study of the bones, Zipfel and I published our findings on these fossils. We detailed the unusual apelike anatomies of the heel, the ankle, and the sole of the foot. But what we couldn't figure out was what it actually meant for how *sediba* walked. Bone for bone, *sediba* was different from other species of *Australopithecus* or humans today. It must have walked differently; we just didn't know how.

Many paleoanthropologists are trained to be experts in one particular part of the skeleton. Among us are experts in the anatomy of the skull, teeth, elbows, shoulders, knees, and hips. My specialization is the foot, and in particular, the ankle. We are trained this way in part because paleoanthropology is a science of fragments. In six weeks at a fossil site, we may find a couple of hominin teeth and, if we are lucky, a hominin elbow here and foot bone there. To make sense of these ancient bits of bone, we learn everything we can about how one bone varies among different kinds of animals, how those differences affect the way an animal lives, and how that bone has evolved throughout ape and human evolution. This is what it takes to squeeze every bit of information we can out of the precious hominin scraps we find.

That is, until we find a nearly complete skeleton. And in the case of *sediba*, we had two.

We are so accustomed to interpreting isolated fossil fragments that, when presented with a skeleton, it is easy to treat it as a bunch of parts. But skeletons are not collections of unconnected anatomies. They are the bony remains of what was a cohesively operating system when the individual was alive. To interpret the odd Malapa skeletons, we needed the advice of someone who thought every day about how the body works as a unit and how a change in one joint impacts the others.

We needed a physical therapist.

"What does the knee look like?" Boston University biomechanist and physical therapist Ken Holt asked me after I had delivered a presentation on the Malapa fossils to his group. Holt, who has since retired from teaching, maintains a physical therapy practice and devotes most of his efforts toward developing a Tony Stark Iron Man–like suit to help stroke victims walk normally again.

"The knee? It is strange," I told him. "It is humanlike in most ways, but it has the highest lateral lip I've ever seen." The lateral lip is a bony retaining wall in the knee that holds the kneecap (patella) in place. Apes do not have this knee morphology. It is something found only in bipedal hominins. But *sediba*'s lateral lip was so large that it was superhumanlike, which was perplexing given the apelike anatomies of the foot.

"That makes sense," Holt replied.

"It does?" I asked. It didn't to me.

We met again after the Q&A, and he explained what a physical therapist saw in these bones. *Australopithecus sediba*, he said, looked to him like a hyperpronator.

With such a small, chimpanzeelike heel, *sediba* could not have walked with a prominent heel strike as humans do today or even as Lucy's species did. Instead, *sediba* walked more like an ape—taking short steps on a flat foot, striking the ground along its outside edge. For every action, there is an opposite and equal reaction, so when *sediba* hit the ground with the outside edge of its foot, the ground hit back and rapidly rotated the foot toward the big toe. Doing this would cause the shinbone to twist inward, rotating the knee.

Some people today walk like this, and it is called hyperpronation. If you are one of these walkers, the outside edge of the sole of your shoe, especially near the heel, wears down quickly. Because the knee jerks inward, hyperpronators are at risk of dislocating their kneecaps, an injury that occurs to about 20,000 Americans each year. It is still possible to walk with a dislocated kneecap, and it can be popped back into place, although it is painful and full recovery can take six weeks.

It sure sounds like a bad idea for an *Australopithecus* to walk in a way that would predispose it to kneecap dislocations. In fact, it sounds like the start of a recipe for a saber-toothed cat cookbook. But *sediba* had evolved an exceptionally large retaining wall for keeping its patella from dislocating. In other words, this species had anatomical solutions to the very problems that hyperpronating humans have today. They were adapted to move this way.

For several months, Holt and I worked through the mechanics of how *sediba* walked around her world. We corresponded regularly with Zipfel in South Africa, who understands how the human foot operates better than anyone I've ever met. During those months, I hyperpronated around the Boston University campus like a *sediba*, using my own body as a way to understand my 2-million-year-old relative. It hurt at times, and I probably became "that guy" to BU students. But walking this way made it clear to me that while I may be a biped, I'm no *sediba*.

Holt, Zipfel, and I tested our hypothesis and found evidence consistent with it throughout *sediba*'s skeleton. For instance, we spotted a perplexing, apelike curve at the base of a bone called the fourth metatarsal in the middle of sediba's foot, which would have made its foot more flexible. We took MRI scans of forty humans and found this same bony shape in a handful of them—the ones who just so happened to hyperpronate.

Amey Y. Zhang, a Dartmouth student interested in the intersection of art and science, 3D-laser-scanned the *Australopithecus sediba* skeleton and used animation software to rig the joints and make it walk.

The animation was tweeted and retweeted and eventually evolutionary biologist Sally Le Page set it to music, ironically syncing an extinct hominin's walk to the Bee Gees' "Stayin' Alive."

Our colleagues generally agree that *sediba* has peculiar anatomies and therefore walked differently from other *Australopithecus* species. But they don't have to believe me; science is not about beliefs. 3D surface scans of the *sediba* fossils have been posted to a free website (www.morphosource.org), and my colleagues anywhere in the world can access them and test our hypothesis for themselves.

From day one, this has been Berger's approach to the extraordinary discovery of *Australopithecus sediba*. Science can proceed only if hypotheses can be tested, and that can happen only if the fossils are accessible to the entire scientific community.

While some have embraced our hyperpronation hypothesis, others have not. Bill Kimbel, a paleoanthropologist at Arizona State University who holds the excavation permits for Hadar, Ethiopia, wrote, "The proposed 'hyperpronation' of the foot and extreme inward rotation of the leg and thigh suggest an ungainly bipedal stride that might have made it into Monty Python's 'Ministry of Silly Walks' sketch." He later suggested that the gait was "pathologically impaired" since there are no clear selective advantages to walking in this manner.

That is why discovering more than one *sediba* was so important. One skeleton can always be dismissed as pathological. But we had two and even a few bones from a third individual. While we based much of our walking hypothesis on the adult female (MH2), clues also came from the other individuals found at Malapa. Could the small heel of the female skeleton be pathological? Probably not, because we also found the heel of the young male skeleton (MH1), and it is nearly identical. Could the ankle joint of the female skeleton be pathological? Probably not, since we found the ankle of a third individual, and it has the same peculiar shape as the female's. The unusual anatomy of the fourth metatarsal from the young boy is consistent with other bones we found of the adult female.

In other words, I think they all walked this way.

Kimbel is right that *sediba*'s walk is a bit ungainly. Why would it hyperpronate? I think it has everything to do with its reliance on the trees.

Sediba was not the only thing found in the depths of Malapa Cave. Berger and his team of paleontologists recovered the fossilized remains of other animals and even found a coprolite—fossil poop. It was whitish in color and contained some flakes of bone. Marion Bamford, the director of the Evolutionary Studies Institute at the University of the Witwatersrand and an expert in fossil wood and ancient ecosystems, dissolved the coprolite in hydrochloric acid and recovered tiny bits of ancient plant parts and even microscopic pollen that matched trees found today in cooler, wetter, high-altitude forests. Two million years ago, *sediba* walked in the woods.

With its long arms and shrugged shoulders, *sediba* was a skilled climber, but not just for safety. Karabo's (MH1) skull is so well preserved that he still has food stuck between his teeth. Amanda Henry, now at Leiden University in the Netherlands, scrapped plaque off MH1's teeth and discovered phytoliths, tiny silicate remains of plant cells that were still preserved from his last few meals of fruit, leaves, and tree bark. Karabo fed in the trees. Additionally, isotopic analysis of tiny tooth fragments revealed that unlike other species of *Australopithecus*, *sediba* did not dine in the grasslands. Instead, it relied heavily on food from forests, as *Ardipithecus* did millions of years earlier.

It made sense to me that if *sediba* was better adapted for life in the trees than Lucy was, it would compromise, or at least modify, how it walked when it moved on the ground from one forest food patch to the next.

Meanwhile, 2-million-year-old fossils continue to be found at Malapa, and they have more to tell us.

At the University of the Witwatersrand, boulder-size chunks of breccia line the metal shelves, waiting their turn to be prepared. Rocks with hominin fossils projecting from them jump the queue, sending zebras and antelopes to the back of the line. Thus, a large block with a beautifully preserved antelope leg was waiting its turn until scientist Justin Mukanku turned it over and spotted the glimmer of a hominin tooth. A CT scan of the block revealed that it contained missing parts of Karabo, the young male, including his lower jaw, backbones, parts of his pelvis, ribs, legs, and feet. When these fossils are freed from the matrix, they will help us refine our understanding of how this *Australopithecus* walked.

A FEW MONTHS after nine-year-old Matthew Berger tripped over a chunk of rock at Malapa and discovered *sediba*, Stephanie Melillo, a paleoanthropologist at the Max Planck Institute for Evolutionary Anthropology in Leipzig, Germany, was prospecting in a new, 3.2- to 3.8-million-year-old fossil site called Woranso-Mille. It was located in Ethiopia, northwest of Hadar, where Lucy had been found. Melillo was part of a team led by Yohannes Haile-Selassie, curator of paleoanthropology at the Cleveland Museum of Natural History, who had discovered the site and had already found fossils there that he attributed to Lucy's species.

Haile-Selassie, like his advisor Tim White before him, thought at the time that the only hominin on the landscape 3.2 to 3.8 million years ago was *Australopithecus afarensis*. This was convenient. It meant that any hominin humerus, foot bone, or skull fragment they found had to be from *afarensis*.

But what Melillo discovered on February 15, 2009, was a shocking realization that Lucy was not alone.

Many fossils are found early in the morning, when the low sun casts shadows across the landscape, caffeine circulates in the bloodstream, and eyes are fresh. By late morning, there is almost always a lull as the blinding sun reflects off the ancient surface and stomachs growl for lunch. This day was no different for Melillo and her dozen colleagues, who had spread out across the badlands of Woranso-Mille in an area called Burtele.

A paleoanthropologist's best digging tool is last season's rain, which washes away sediment and gently exposes buried bone. Melillo slowly walked up a water-carved gully where the silty sediments graded into a reddish sandstone and spotted a small fragment of bone about the size of a paper clip.

"It is so exhilarating to walk and walk and find nothing but dirt but then all of a sudden spot a fossil," Melillo told me during a Skype conversation. "It is amazing, at that second of recognition, how it just pops."

She put an orange flag in the ground to mark the position of the fossil and carefully picked it up. She could tell it was the base of a fourth metatarsal—one of the bones in the middle of the foot—but it was too incomplete to know if it was from a primate like us or from a carnivore. She strode slowly toward Haile-Selassie, examining the bone from all angles as she walked. He knew this walk. It meant she was bringing a possible hominin to him for inspection.

Haile-Selassie is a fossil magnet. As a graduate student in 1994, he discovered the first fossils of the 4.4-million-year-old *Ardipithecus ramidus* skeleton—two palm bones projecting out of the ancient hillside as though Ardi were reaching her hand out for the paleoanthropologists to grab. A few years later, he found a 2.5-million-year-old skull of a new species of *Australopithecus* called *garhi*, and just weeks after that, he discovered fossils that he would use to name a new species himself: the 5.5-million-year-old *Ardipithecus kadabba*.

At the Woranso-Mille site, Haile-Selassie and his team found a 3.6-million-year-old partial skeleton from Lucy's species that they nicknamed Kadanuumuu, an Afar word that means "big man." He has since discovered a new species of *Australopithecus*, called *deyiremeda*, and in 2019 he found the oldest skull of *Australopithecus*—a marvelous 3.8-million-year-old cranium from the species *anamensis*.

"He has a spidey-sense where to find fossils and when to keep digging," Melillo said.

After he found half of an *Australopithecus* jaw on the surface of the ground, she remembered, his team scraped and sieved the area for a week. Scraping and sieving is laborious, boring, and painstakingly slow. Every pebble and chunk of dirt must be closely examined.

"I bet Yohannes we wouldn't find the other half of the mandible," Melillo recalled. "I lost."

One day in the field, Haile-Selassie picked up a fossilized long bone shaft. Long bone shafts are common but often difficult to identify. They could be shards of arm or leg bones from any of dozens of different animals from hominins to antelopes.

"Look at this!" he said excitedly, but Melillo was unimpressed. Weeks later, in the hominin vault at the National Museum in Addis Ababa, he slid the fragment onto the broken surface of a hominin humerus that had been discovered ten years earlier. Somehow, when Haile-Selassie had plucked this fossil from the ground, he knew that its broken surface would fit perfectly onto a fragmentary arm bone that he hadn't seen in a decade.

At the Burtele site, the first thing Haile-Selassie noticed about the foot bone Melillo handed him was that the break in the bone was clean. That meant the fossil had broken recently and that the other half should be nearby.

"Where is the other half?" he asked her.

It wasn't long before team member Kampiro Kayranto found it. The new piece lacked the distinctive ridge found on carnivore foot bones. It was a hominin fossil—only the third intact fourth metatarsal from this time period ever found.

It meant the crawl was about to begin.

The team of around fifteen gathered at the base of the gully, formed a shoulder-to-shoulder line, dropped to their hands and knees, and crept along on the hard ground, picking up every tiny piece of bone they could find. First, they found a couple of hominin toe bones. When they reached the reddish sandstone layer, they discovered the big toe and the second digit poking out of the ancient soil.

These weren't just individual bones. They were piecing together a partial foot skeleton. By dating the volcanic ash that sandwiched the reddish sandstone, they determined that the fossils were around 3.4 million years old.

It had been known for decades that Lucy's species, *Australopithecus afarensis*, lived in these parts at that time. But this was no *afarensis* foot.

It was more like an *Ardipithecus*, the big toe and the second toe facing one another like a human thumb and index finger. The Burtele foot belonged to a more apelike hominin—something that climbed more often in the trees and walked on two legs differently than Lucy did.

Back in the lab at the Cleveland Museum of Natural History, Haile-Selassie showed the bones to foot expert Bruce Latimer, who was part of the team that described Lucy in the early 1980s and Ardi in the 2000s.

"You found another Ardi!" Latimer recalled saying when he first saw the fossils, certain that the foot was from the 4.4-million-year-old hominin. "Great!"

No, Haile-Selassie told him. The bones were found in sediments a *million* years younger.

"Can't be," Latimer replied with astonishment.

The Burtele foot had some of the key anatomies for bipedal walking, but it was clearly different from the foot that made the Laetoli G-footprints. In this prehistoric version of the Cinderella story, the Burtele foot would not fit the Laetoli slipper.

Subsequent discoveries confirmed the presence of a new species: *Australopithecus deyiremeda*. Another hominin, walking in a different way than Lucy's species, coexisted with her and her kind. In fact, in 2021 we concluded that the Laetoli A-footprints were made by a second species of hominin. Recall that these prints were so different from the Site G tracks that some researchers thought they had been made by a bear. Instead, we now think two species walked through the muddy ash at Laetoli 3.66 million years ago.

We used to think that throughout human evolution, there was only one way to walk. But we now know that is not the case. Millions of years ago, different yet related species of *Australopithecus*, living in different environments, walked in slightly different ways. They trod over much of Africa, establishing a huge range from north-central grasslands and down the eastern Rift Valley from Ethiopia to South Africa—a distance of nearly 4,000 miles.

About 2 million years ago, members of our own genus, *Homo*, evolved. They had slightly smaller teeth, larger brains, and a greater proclivity for stone tool use. What remains a mystery is which *Australopithecus* species evolved into *Homo*. It could be one we haven't found yet.

By 2 million years ago, bipedalism had become a grand evolutionary experiment—one about to take its show on the road.

CHAPTER 8

Hominins on the Move

There was nowhere to go but everywhere,
so just keep on rolling under the stars.

—*Jack Kerouac,* On the Road, *1957*

In 1983, archaeologists were excavating at the medieval site of Dmanisi in the country of Georgia, which was then part of the Soviet Union. The archaeological team was finding coins and other medieval artifacts when they came across a tooth. Thinking it was probably from an animal that had been eaten by traders who stopped at Dmanisi as they traveled the Silk Road, they brought the tooth to Abesalom Vekua, a trained paleontologist. This, he determined, was no cow or pig. The tooth belonged to a rhinoceros.

What was a rhinoceros doing in a medieval grain pit in this mountainous section of Southwest Asia? Vekua and his colleague Leo Gabunia decided to investigate the source of their out-of-place rhinoceros.

One clue was that the tooth wasn't modern. It was from *Dicerorhinus etruscus*, a species that went extinct in the Pleistocene. The following year, they dug at Dmanisi and discovered stone tools similar to the simple Oldowan ones Mary and Louis Leakey had found at Olduvai Gorge in Tanzania. The mystery of the rhinoceros tooth began to make sense. It turns out that the Dmanisi citadel was built on Pleistocene sediments. The archaeologists digging through the dirt for medieval relics had penetrated through to a much older layer, a distant time when *Dicerorhinus* roamed the landscape.

It was also a time when hominins were not supposed to have expanded their ranges beyond the borders of Africa. But those stone tools indicated that they had.

Vekua and Gabunia kept digging, and in 1991, they found a hominin jaw. A decade later, they unearthed two skulls from sediments resting on a 1.8-million-year-old lava bed. The skulls had large faces, but their brains were about half the size of human brains today. They were identified as coming from an early version of *Homo erectus*, a species that had been known to science since Eugène Dubois's discoveries in the late nineteenth century. In the two decades since, three more skulls and two partial skeletons have been unearthed from this amazing site. The Dmanisi hominins are the oldest ever discovered outside the African continent.

However, evidence from the other end of the Silk Road, at Shangchen in central China, indicates that hominins were on the move even earlier.

In 2018, Zhaoyu Zhu of the Guangzhou Institute of Geochemistry at the Chinese Academy of Sciences announced the discovery of simple stone tools made by ancient human hands 2.1 million years ago. About the time *Australopithecus sediba* was hyperpronating around South Africa, another branch of the human family tree had pushed almost 9,000 miles to the east. No bones have been found yet, so we don't know who made these stone tools, but most paleoanthropologists assume early *Homo erectus*—or perhaps an even earlier representative of our genus.

At first glance, the spread of ancient humans around the world seems sudden. Hominins had been in eastern and southern Africa for millions of years, but now, seemingly in a flash, they were in

China. However, this was not as rapid as it appears. If early members of the genus *Homo* migrated east at just one mile per decade starting around 2.2 million years ago, they could have reached China 2.1 million years ago, in plenty of time to leave their stone tools at Shangchen.

The discoveries at Dmanisi and Shangchen reveal that soon after *Homo* evolved in Africa about 2.5 million years ago, their territories expanded, spreading north and east into Eurasia. There was no WELCOME TO ASIA sign greeting them. They didn't *know* they were moving into regions that would amaze and puzzle their descendants 2 million years later. But this does raise some questions.

Why did hominins become explorers at this time? And how were they able to move into territories that had been uninhabited by their ancestor *Australopithecus*?

The clues can be found in the skeleton of a boy.

IN 2007, I traveled to Nairobi, Kenya, a densely populated city 6,000 feet above sea level. For two weeks in August, the weather was surprisingly cold and cloudy. It didn't rain, but the air was heavy and still. The streets were lined with vendors selling fresh fruit and nuts. Goats roamed, eating roadside litter. Smoldering piles of trash added to the unpleasant smell of diesel fuel. The day I arrived in Nairobi, a head cold materialized and pressed hard against my sinuses for a week.

Nairobi is a city of over 3 million people, though that number rises to over 6 million if the surrounding population is counted. This includes the largest slum in Africa—Kibera, where nearly 1 million people live on an average income of less than $1 a day. A few miles north of Kibera, atop Museum Hill in the Westlands district, sits the Nairobi National Museum, where some of the most prized fossils ever unearthed are housed in a vault the size of a small coffee shop.

Outside the museum stand statues of Louis Leakey and a large orange dinosaur. I skirted past the public exhibits and through a courtyard to the research collections, where I met Fredrick Manthi, a Kenyan paleontologist who often goes by his middle name, Kyalo.

Manthi's father worked on Mary Leakey's expeditions in the 1970s, and young Kyalo caught the hominin bug at an early age. After earning his Ph.D. at the University of Cape Town, he returned to Kenya to lead the paleontology and paleoanthropology division at the museum and thus oversees all prehistory research in Kenya. Three years after I met him, he would discover a beautiful 1.5-million-year-old *Homo erectus* skull near the village of Ileret on the east side of Lake Turkana.

I gave Manthi the list of fossils I wanted to study, ranging from the 20-million-year-old foot bones of the ancient ape *Proconsul* to a fossil femur from an archaic *Homo sapiens*. I figured that on my first day he would bring me a tray of fragmentary foot fossils, the kind of bones only a handful of people in the world cared about. I was still a student, after all, and was groggy from the cold medicine.

Instead, Manthi reemerged from behind the thick steel door of the vault with a wooden tray containing the Nariokotome *Homo erectus* skeleton. It was like giving the curator of the Louvre a list of Renaissance paintings to study and being handed the *Mona Lisa*. My arms went weak and my hands trembled. Manthi may have noticed my open jaw as I ogled the skeleton. Instead of handing me the tray, he walked them to where I would be working and carefully rested the precious fossils on the counter.

I love fossils. I travel far to see them, eager to take measurements, photographs, and 3D scans of these fragile fragments of our past. But for the first few minutes of every visit with a new fossil, my calipers, camera, and scanner remain idle. I just sit, alone, with the remains of my ancestors. I appreciate the color, texture, and curve of every piece. I wonder not only about the species but also about the *individual* whose death and preservation allows us to understand our own place in the story of life. I let myself be moved. I let myself be emotional. This ritual started in August 2007, when I sat, alone, in the Nairobi National Museum with the Nariokotome skeleton.

Then I got to work.

The Nariokotome fossil was discovered in 1984 by Kamoya Kimeu, arguably the most prolific hominin discoverer in history. He was part of the Leakey family's famous "hominid gang," whose discoveries in Eastern Africa in the 1960s and 1970s opened the floodgates for paleoanthropological research in Tanzania, Kenya, and eventually Ethiopia. Kimeu's discoveries are so important that two fossil species—a Miocene ape, *Kamoyapithecus,* and the early Pleistocene monkey *Cercopithecoides kimeui*—have been named after him.

In their book *The Wisdom of the Bones*, paleoanthropologists Alan Walker and Pat Shipman described Kimeu's approach to fossil hunting as "walk, and walk, and walk, and *look* while you are doing it."

On August 22, 1984, he was doing just that on the west side of Lake Turkana. Along the bank of the dried up Nariokotome River, he spotted a tiny fragment of skull bone, camouflaged the same dark color as the surrounding sediment.

"Lord knows how he saw it," Walker and Shipman wrote.

Kimeu called Richard Leakey and Alan Walker in Nairobi, and the project leaders arrived the next day. For the next five years, the team moved 1,500 cubic meters of dirt. Hidden within it was most of the skeleton of a juvenile *Homo erectus* who died 1.49 million years ago.

The Nariokotome Boy, as he came to be known, is one of the most complete and important skeletons ever discovered. He reveals the kind of body it took for early *Homo* to expand its range beyond the borders of Africa.

His brain, which had reached full size, was only two-thirds the volume of a modern human's. The unerupted wisdom teeth and the unfused growth plates in his arms and legs tell us that he was young when he died—just nine years old according to a detailed study of his teeth. But his leg bones also indicate that he was already over five feet tall and weighed close to one hundred pounds. That is a big kid. My son was almost a foot shorter and forty pounds lighter at that age.

Scientists calculate that the Nariokotome child would likely have been close to six feet tall had he survived into adulthood. The boy's large size at such a young age also indicates that his species hadn't evolved the adolescent growth spurt we have today. Why not? Northwestern University anthropologist Chris Kuzawa discovered a trade-off in energy allocation between the brain and body in children. Kids' brains use so much energy during their preteen years that the growth of their bodies slows down. In the teen years, bodies play catch-up and rapidly grow in height—the growth spurt. Because the brains of *Homo erectus* were only two-thirds the size of ours, they probably could still divide energy between the brain and a growing body.

More about this species can be gleaned from a fragmentary skeleton of a 1.6-million-year-old adult *Homo erectus*, a specimen named KNM-ER 1808. It was discovered in 1973 by, of course, Kamoya Kimeu. Its large right femur is the size of a thigh bone in a modern human standing a shade under six feet. People tend to think that modern human sizes were not achieved until recently, but that is wrong. *Homo erectus* was well within the size range of people today.

I turned back to the tray containing the Nariokotome skeleton and plucked his left femur from its aqua-colored foam bed. The femur is dark gray, with splotches of black and brown. I was struck by its length. He also had a large upper arm bone (humerus) 34 percent longer than Lucy's. That makes sense because he was a bigger individual than Lucy, but you might expect, then, that his femur would also be 34 percent longer than hers. It wasn't. It was *54* percent longer.

Homo erectus was not a scaled-up *Australopithecus*. Legs had gotten longer.

"From ants to elephants, the variable that explains how much energy an animal uses to get from one place to another is leg length," Herman Pontzer, a professor of anthropology at Duke University, told me. As legs get longer, his extensive research shows, travel generally becomes easier.

Equipped with longer legs, our *Homo erectus* ancestors could range farther than Lucy's kind. But that's not all. *Homo erectus*, it turns out, had also evolved a foot with a fully modern human arch.

In 2009, a team of researchers from the Nairobi National Museum and George Washington University discovered nearly one hundred fossil footprints near Ileret. They were made along the muddy shores of a lake by twenty *Homo erectus* individuals 1.5 million years ago. They are the size of human footprints today and feature a prominent arch that put a spring in their step—especially when they ran.

Australopithecus species had an arch, too, but it was low by modern-human standards. In *Homo erectus*, the arch was fully modern, legs were long, and our ancestors finally had the anatomical equipment to range farther and collect more food.

In ecosystems throughout the world, carnivores have, on average, larger home ranges than herbivores. Plants often grow in clumps, so herbivores do not need to range as far each day to find food. Carnivores, however, have to search far and wide to hunt down a meal. It is no coincidence, then, that *Homo erectus* fossil sites contain a lot of butchered animal bones acquired both by hunting and scavenging.

Stone tools date back to 3.3 million years ago, and cut marks are found on 3.4-million-year-old antelope bones that predate *Homo erectus*. It seems, then, that *Australopithecus* and even early *Homo* dabbled in carnivory through opportunistic scavenging. But they were not hunters. With *Homo erectus*, scavenging became more prevalent, and there is even evidence of deliberate, coordinated hunting. Plants remained part of their diet as well, making them omnivores just as we are. Of course, some people today choose not to eat meat, but there is ample evidence that meat and marrow were important resources that helped our lineage survive the Pleistocene.

With long legs, arched feet, and expanded home ranges, *Homo erectus* pushed beyond the borders of Africa and into Eurasia.

THE *HOMO ERECTUS* skeletons at Dmanisi, Georgia, are not as tall as the Nariokotome Boy. They are barely over five feet. But they have long legs and the body proportions of modern humans. The Dmanisi hominins could walk with great efficiency and follow game through the Middle East and modern-day Turkey into the Caucasus. Even earlier migrations had made it all the way to China. It remains unclear whether the hominins followed the mainland across the plateaus of Asia or if they followed the coastline through India and Southeast Asia. Either way, 2.1 million years ago they made it clear across the largest continent on Earth.

Human migration narratives are often presented in an oversimplistic and unidirectional way, but the odds that these territorial expansions happened only once and in only one direction are infinitesimally small. *Homo erectus* almost certainly moved in and out of Africa in pulses, gradually exploring the edges of their expanding range as they ventured into territories never before inhabited by an upright walking hominin. By at least 1.5 million years ago, *Homo erectus* had pushed as far southeast as they could walk without getting their feet wet.

During ice ages, at least eight of which happened cyclically during the last million years, enough water was trapped at the poles and in mountain glaciers to make sea levels drop, so it would have been possible to walk from Southeast Asia to Java in the Indonesian Archipelago. But no farther. There, *Homo erectus* would have encountered a twenty-mile-wide, five-mile-deep oceanic trench that delineates Wallace's Line, named for the nineteenth-century naturalist Alfred Russel Wallace, codiscoverer with Charles Darwin of natural selection. To the west of Wallace's Line are plants and animals found in Asia; to the east, the startlingly different plants and animals of Australia. This ecological boundary is nearly impossible to cross without a boat.

Around the time *Homo erectus* reached Java, hominins were also spreading into western Eurasia. In 2013, Spanish paleoanthropologists announced the discovery of a single hominin tooth and stone tools from a cave in the town of Orce in southeastern Spain. They were deposited in 1.4-million-year-old sediments. A few years earlier, a more complete 1.2-million-year-old lower jaw was found by Eudald Carbonell at a cave called Sima del Elefante, meaning pit of the elephant, in the Atapuerca region of northern Spain. The researchers called the fossil *Homo antecessor*, which means "pioneer."

Homo erectus and its cousins had become cosmopolitan apes, ranging from the tip of South Africa to as far west as Spain and as far east as Indonesia. There were no wagons, planes, trains, or automobiles. No domesticated horses to ride. They walked.

During that time, a strange and wonderful thing was happening. Brains were getting bigger. A lot bigger. Two nonmutually exclusive hypotheses explain why this might have been happening, and they both have to do with food.

The first, formulated by anthropologists Leslie Aiello and Peter Wheeler in 1995, is called the expensive-tissue hypothesis. Aiello and Wheeler collected data on organ weights of primates and reported that humans are unusual in having a very large brain (which everyone knew) but an extremely short gut (which everyone did not know). Guts are energetically expensive to maintain, constantly sloughing off old tissues and regenerating new ones. By evolving a shorter gut, these hominins had ostensibly freed up energy that could be reallocated to brain growth. This would not work in a strict herbivore. Plant eaters require long hindguts to digest tough cellulose fibers in plants. Carnivores, in contrast, have short hindguts that absorb nutrients from meat and marrow without meters and meters of intestines. Aiello and Wheeler proposed that as our ancestors consumed more animals, those with shorter guts and larger brains thrived and multiplied. Between 2 million and 1 million years ago, the average hominin brain roughly doubled in volume.

More recently, Richard Wrangham, a human evolutionary biologist at Harvard University, introduced another variable into the equation: fire.

Tantalizing evidence from the east side of Lake Turkana in Kenya and from Swartkrans Cave in South Africa indicates that by 1.5 million years ago, *Homo erectus* had learned how to control fire. The 1-million-year-old South African cave Wonderwerk has incontrovertible evidence of fire use. With it, our ancestors could cook their food, making it easier to digest. This, Wrangham suggests, provided them with the energy they needed to evolve larger brains. Fire also would have allowed our predecessors to spread into territories previously too cold to inhabit. And it would have freed them from escaping to the trees for safety at night since fire is a predator deterrent. As *Homo erectus* evolved longer legs, they became better walkers, but climbing became difficult. With fire, these disadvantaged climbers could survive and multiply.

And as our ancestors walked, they began to talk. As the phrase goes, "If you're going to talk the talk, you've got to walk the walk." It turns out, walking and talking are indeed linked.

In four-legged animals, the muscles of the shoulders, chest, and even abdomen absorb the impact as their front limbs hit the ground. This means that animals moving on all fours have to coordinate their breathing and walking, one breath for each step. That's why animals can't gallop and pant at the same time. Such short, quick breaths are not possible when digestive organs slam up against the diaphragm with every stride. Because they can't pant, most running animals are unable to cool down, so they have to stop and rest in the shade after short sprints. Humans, however, can breathe rapidly as they stride. Unlike many four-legged animals, we also sweat. This allows us to cool off while running. We are slow, but we can go for miles.

But what does this have to do with language?

You can mimic the role of chest and arm muscles in a quadruped by carrying something heavy while you walk. Your chest muscles tense and you draw a breath with each stride. Aside from the occasional grunt, it's difficult for you to make sounds. That is what it is like for a four-legged animal. But animals that walk on two feet have finer control of their breathing, and that gives them the flexibility to make a great range of sounds.

Geladas, terrestrial monkeys that live in the Ethiopian highlands, sit upright while they graze on seeds. In a seated position, they communicate through a series of complex vocalizations. As bipeds, humans have evolved language, combining sounds produced through fine muscle control of our breathing into a seemingly infinite number of combinations and meanings. Even in our children, the onset of walking and talking are closely linked.

The origin of human language remains unknown and controversial. Many factors beyond breathing flexibility help us produce sounds. The base of our skull and the vocal apparatus in the back of our throats form a resonance chamber absent in the apes. Our hyoid, a bone rarely found in the human fossil record, is thick to anchor muscles and ligaments used when we speak. Regions of the brain such as Broca's and Wernicke's areas are critical for language production and comprehension. Our inner ear bones are fine-tuned to the frequencies of human voices.

Early on, hand signals may have been as important as the spoken word in the development of language. The relationship between sounds and meaning may have started with onomatopoeias—words that sound like what they mean. Birds chirp. Bees buzz. Hands clap. But onomatopoeia doesn't work for everything. What does a hunt or a sunrise sound like? So some sounds were needed to symbolically represent meanings. We had to become symbolic apes. Eventually, songs and music also played a role in spreading ideas and preserving memories. These pieces did not emerge all at once, but we can try to reconstruct when the first languages were developing in our ancestors by picking through the fossil record.

Bipedalism likely gave *Australopithecus* the fine-controlled breathing required to make a larger range of sounds than a chimpanzee can make, and it freed their hands to communicate with gestures. But there is little evidence that they actually talked.

A hyoid bone from the 3.4-million-year-old Dikika Child looks like an ape's. Fossilized brain impressions and CT scans of the inside of fossil skulls reveal that the folds and fissures of the earliest *Australopithecus* brains were quite apelike. But in some *Australopithecus*, there appears to be asymmetry in the Broca region, suggesting that the brain was primed for producing and understanding language. Certainly, this was the case for early *Homo* brains by around 2 million years ago.

Half-million-year-old fossils from Spain reveal both humanlike hyoid bones and inner ears fine-tuned to detect and process sounds in the vocal bandwidth. And genetic evidence indicates language may have been present by this time. DNA extracted from fossil hominins in Europe and Asia shows that a gene that impacts language, although exactly how it does that is unknown, evolved into its current form by at least 1 million years ago.

All of the key ingredients for language appear to have been in place by half a million years ago, but the first step in this evolutionary sequence was upright walking, which provided the fine control of breathing required to make a large repertoire of sounds.

Homo erectus walked, and as it spread through the world, it also talked.

THE COMING AND going of ice ages through the Pleistocene at times permitted hominins to reach places that would otherwise be inaccessible, only to then trap and isolate them. *Homo erectus* individuals living on what is now the island of Java, for example, could wander throughout southeastern Asia during a glacial maximum but then were stuck on the island for tens of thousands of years when the seas rose during a warm spell. In Western Europe, glacial periods permitted hominins to even reach England. We know this because they left footprints there.

Around 800,000 years ago, a group of hominins sometimes given the species name *Homo heidelbergensis* walked along a muddy shore near modern Happisburgh, England, leaving footprints nearly identical to yours and mine. The shoreline is eroding fast, however. Soon after researchers photographed and measured the prints, they were washed away. The Pleistocene population that made the footprints was also fleeting, eventually pushed south and isolated in pockets along the Mediterranean as glaciers advanced from the north.

These climatic pulses led to the intermittent genetic isolation of Pleistocene *Homo* populations. One of them, initially isolated in pockets in Europe and western Asia, evolved into the Neandertals. Their bones are plentiful and have been known to science since the mid-nineteenth century. Once glaciers retreated, they expanded their range from Portugal to Ukraine. Two dozen complete skulls have been unearthed.

In 2019, scientists from the National Museum of Natural History in Paris announced the discovery of an astounding collection of 257 Neandertal footprints made in the dunes of Normandy, France, 80,000 years ago. A dozen children walked with an adult or two through the wet sand, immortalizing a day in the life of a Pleistocene day care. I imagine the Neandertal kids playing and laughing while the adults scanned the horizon for threats.

At this time, parts of Asia were inhabited by the Denisovans, a group known not just from the anatomy of a handful of scarce fossils, but from DNA extracted from tiny scraps of bone found in Siberian and central China caves.

Homo erectus and its cousins, with their long legs, enlarged brain, and control of fire, had spread throughout Africa, Asia, and Europe. The stage was now set for the final phase of our journey—*Homo sapiens*.

But recent discoveries have disrupted this narrative, hinting that human evolution and the migration of hominins around the globe were much more complicated, and even more interesting, than any of us imagined.

Except, maybe, J. R. R. Tolkien.

CHAPTER 9

Migration to Middle Earth

Not all those who wander are lost.

—*J. R. R. Tolkien,* Lord of the Rings: The Fellowship of the Ring, *1954*

Each autumn, leaf peepers flock to northern New England to see what happens when the sun's rays are low in the sky and the sugar maples, birches, and oaks stop producing chlorophyll. Brilliant reds, oranges, and yellows paint the hillsides as the days shorten and the air grows crisp.

Applebrook Bed & Breakfast in Jefferson, New Hampshire, provides one of the best views of nature's palette. Jefferson lies at the mouth of a pass through the White Mountains—Mts. Waumbek and Cabot to the north, the Presidential Mts. Adams, Jefferson, and Washington to the south.

"It is hard to go west to east in northern New England," Dartmouth College archaeologist Nathaniel Kitchel said as we drove in that direction along Route 2 from Vermont toward Jefferson. The mountain ranges in Vermont and New Hampshire form north-to-south walls, and the snowy east-west passes are often closed for months in the winter and spring.

The few east-west paved roads here lie atop dirt roads that cover horse trails, footpaths, and game trails first trod by the stomping feet of Pleistocene mammoths and mastodons.

"The earliest inhabitants to this area would have followed this same route," Kitchel said.

Archaeologists call it "paleo-route 2." Along here, 12,800 years ago, humans began moving into uninhabited lands.

By that time, the Laurentide Ice Sheet, once thick enough to cover the 6,288-foot Mt. Washington, had retreated, carving valleys and dumping large glacial boulders in its wake. The melting edge of the ice created thousands of lakes, including one a half-mile wide at Jefferson, where all that remains of it now is the gentle Israel River. Large caribou herds shared the landscape with woolly mammoths and beavers the size of Saint Bernards. The maples, birches, and oaks had not yet moved north to cover the granite hills.

It was an unimaginable scene—a New England without trees. But even then, the spot where Applewood Bed & Breakfast now sits was the best view in town.

Kitchel and I traveled there on a cold December day. The sky was a brilliant blue brushed with the occasional feathery cirrus cloud. At noon, the winter sun was low in the sky, making it feel later in the day than it was. The tops of the mountains were frosted with snow and ice. It was a good time of year to imagine the cold, treeless scene in a period of the Late Pleistocene called the Younger Dryas, when humans first came through here. The hills looked bare. If I squinted my eyes, I could imagine that there were no trees at all. The golf course in the foreground helped.

In 1995, a storm blew down a tree behind the Applewood B&B. Though not an unusual event by any means, Kitchel reminded me, "archaeologists are always looking at the ground." Local resident and amateur archaeologist Paul Bock examined the base of the uprooted tree and spotted a stone tool. It was a fluted point—a specific kind of tool made by some of the first people in the Americas. Dick Boisvert, the New Hampshire state archaeologist, excavated the area with teams of students for two decades and found evidence that, starting just under 13,000 years ago, humans routinely camped at this spot.

From this vantage point in Jefferson, with no trees blocking their view, the first people to this area could see straight across the valley and spot herds of caribou, a prowling dire wolf, or the smoke from their neighbor's campfire. Sometimes other humans would pass through, but they were no threat. There was plenty of food—caribou, cattails, wetland tubers—and no need, nor any archaeological evidence, for hostility. It was cold, but these people were the descendants of populations that had emigrated from Siberia into the Americas a few thousand years earlier. They knew how to survive in this climate. They had invented bone needles and could stitch warm clothing and waterproof shoes from the hides of the animals they hunted.

New England brings to mind maple syrup, dropped *r*'s, and Super Bowl rings, but not archaeology. That doesn't deter Kitchel, who grew up in northern Vermont.

"In many ways," he says, "the first folks to step foot in New England represent the last pulse of pedestrian-powered human settlement of uninhabited lands—the culmination of a process that began in Africa millennia before."

He's right. But to understand how humans eventually spread all the way to Jefferson, we have to travel back to Africa, 300,000 years ago.

SIMPLE NARRATIVES IDENTIFYING the specific time and place of our species' origin are alluring but wrong. In 2019, for example, a study published in a journal whose standards could be higher boldly declared that all modern humans originated in the northern corner of Botswana in southern Africa. Such claims ignore the obvious. Humans move, and we always have.

The earliest fossil record of *Homo sapiens* demonstrates this fact. The three oldest fossil skulls discovered from our species have been found in Morocco, South Africa, and Ethiopia—the geographic corners of the enormous triangular-shaped African continent. We did not evolve in one specific place at one precise time. Instead, our species slowly evolved as hominin populations moved throughout Africa and exchanged genes, some favorable for survival.

Recent studies examining entire genomes of humans, both past and present, time this pan-African evolution of *Homo sapiens* to between 260,000 and 350,000 years ago. That is not to say that the origin of our species happened at some specific moment between those dates. Rather, our species gradually evolved over that entire period as we walked all over the continent.

The site of Olorgesailie in Kenya helps us understand what happened. Olorgesailie is where I first cut my teeth as a paleoanthropology student in 2005. Badlands in the shadow of Mt. Olorgesailie feature alternating bands of lakeshore sediments, ancient soils, and volcanic ash layers. Fossils from ancient elephants, rhinoceroses, and extinct baboons the size of small gorillas erode out of the barren hillsides. Everywhere you look, there are stone tools. Human ancestors were obviously here, but oddly, of over 70,000 fossils collected at Olorgesailie, only two—a skull fragment and a jaw—are from hominins. Early humans lived at Olorgesailie, but they didn't die there.

Smithsonian Institution scientists Alison Brooks and Rick Potts have been working at this site for decades, and in 2018 they discovered obsidian stone tools buried in 300,000-year-old sediments. The obsidian didn't come from nearby. It matched the chemistry of the rocks in quarries located sixty miles away. They also found black manganese and red iron-rich rocks that would have been ground into a powder, mixed with fat, and used as body paint.

At Olorgesailie, at the dawn of *Homo sapiens*, our ancestors were thinking symbolically and exchanging ideas and goods across large distances. We are explorers. We are travelers. We walk. And those walks took us into new lands.

In 2019, Katerina Harvati of the University of Tübingen announced the discovery of two fossils in a cave in Greece. The first, a 170,000-year-old Neandertal skull, was expected. The second fossil was not. She found the back of a 210,000-year-old skull shaped just like a *Homo sapiens*. A year earlier, scientists announced the discovery of a 190,000-year-old *Homo sapiens* upper jaw found by students on their first archaeological dig in a cave near Mt. Carmel, Israel.

It appears, then, that our species expanded into the Middle East and Eurasia earlier than once thought, perhaps only to be pushed back by the occupants of those lands, the Neandertals. This likely happened over and over again in a dynamic process not captured by any static map with arrows. By around 70,000 years ago, however, the dam broke, and *Homo sapiens* flooded into Europe and Asia.

We know from DNA miraculously still preserved and meticulously sequenced by Max Planck Institute for Evolutionary Anthropology scientist Svante Pääbo that *Homo sapiens* interbred with the Neandertals and Denisovans, absorbing their gene pool into our own. Traces of these extinct populations can still be found in our DNA today.

We walked as far southeast as one can go, right to the edge of the Indonesian Archipelago. There, we stood looking across miles of water as our *Homo erectus* ancestors had done before. Perhaps we saw thin filaments of smoke rising from the horizon, the product of natural brushfires burning in the distance. Perhaps we wondered if there were people like us over there. Instead of turning around, some of us built boats and went into the unknown.

We were on the Australian mainland by 65,000 years ago. By 20,000 years ago, we had walked clear across that continent to the southeastern region, where we left dozens of footprints in muddy sediments around the Willandra Lakes.

Others walked north. With warm, waterproof clothing and the control of fire, we trod over snow and ice, through Arctic tundra, and settled on a large swath of land that then connected Asia to North America. These populations prospered and eventually continued east into the Americas.

To traverse these wild, frigid landscapes would have required an important technological innovation: shoes.

IN THE PACIFIC Northwest of the United States, a string of active volcanoes including Mt. Rainier, Mt. St. Helens, and Mt. Hood tower between 10,000 and 14,000 feet above sea level. Their sister, 12,000-foot Mt. Mazama, dominated the landscape of southern Oregon until 7,700 years ago, when it erupted so violently that it collapsed, creating a 4,000-foot-deep, 6-mile-wide caldera that slowly filled with rain and glacial meltwater. Today, it is the deepest, clearest, and cleanest lake in the United States, known by the local Klamath tribe as Giiwas and by the United States National Park Service as Crater Lake.

Just fifty miles northeast of it is Fort Rock, a cave with a long-documented history of occupation by the First Peoples of the Americas. In 1938, anthropologist Luther Cressman, who was married for a time to Margaret Mead, excavated Fort Rock. Under a thick layer of volcanic ash deposited by the Mazama eruption, he made an extraordinary discovery—the remains of seventy-five sandals. They were made from sagebrush bark, peeled and twisted together in a pattern that resembles a flat wicker basket. The forefoot would have slipped into the front of the sandal, which had straps that fastened around the back.

Carbon dating, possible for organic materials less than 50,000 years old, revealed that the sandals were made roughly 9,000 years ago.

The Fort Rock Cave sandals are the oldest shoes ever discovered, but perishable materials such as sagebrush bark are rarely preserved in the archaeological record. Humans were wearing shoes long before Mt. Mazama erupted. To explore when our ancestors first shod their feet, we must rely on other lines of evidence.

Erik Trinkaus of Washington University in St. Louis is an expert in Late Pleistocene human evolution, specializing in the last quarter of a million years of our lineage's history. In 2005, he discovered that humans once had thicker toe bones. His explanation for why our toes became thinner and weaker is that we started wearing shoes. Once we covered our feet to protect them, the bones of our toes did not grow as thick as we grew into adulthood.

The oldest fossil site preserving a human skeleton with thinly developed toes is Tianyuan Cave, located just outside of Beijing, China. It is 40,000 years old.

Foot bones consistent with regular shoe wearing were also discovered at Sunghir, a 34,000-year-old site over a hundred miles east of Moscow, Russia. There, scientists have excavated several deliberately buried skeletons decorated with thousands of mammoth ivory beads. The burial site is located at 56 degrees north latitude, parallel to Sweden, Alaska, and Hudson Bay, Canada—really cold places where shoe wearing was essential to prevent frostbite.

Thirteen thousand years ago, several descendants of the people who crossed the land bridge from Asia to America walked along the shore of Calvert Island in British Columbia, Canada, leaving twenty-nine footprints. The First Peoples of the Americas continued south and made it all the way to Chile by 12,000+ years ago. Around that time, the people who trekked east in their moccasins made it to what is today New England, and someone lost or discarded a fluted point on a ridge overlooking a beautiful valley in Jefferson, New Hampshire.

From 70,000 to about 10,000 years ago, *Homo sapiens* walked until we populated the globe. Along the way, however, we discovered that we were not alone.

IN 2003, A team of Australian and Indonesian scientists were digging in a cave called Liang Bua on Flores, an island in the eastern part of Indonesia. For years they had been finding stone tools that they assumed had been made by *Homo sapiens*. They were, after all, east of Wallace's Line, and the sediments were only about 50,000 years old.

On the morning of September 2, Benyamin Tarus climbed down almost twenty feet into a pit to resume his layer-by-layer excavation, continuing the work his father had begun thirty years before. Digging through a clay layer, he exposed the top of a skull. Indonesian archaeologists Wahyu Saptomo and Rokus Due Awe identified it as a human but, given its small size, they agreed it had to be the partial remains of a child.

When they cleared the clay and dirt covering the teeth, they were shocked. The wisdom tooth was erupted and worn. This skull was from a full-grown adult with a brain barely larger than a chimpanzee's.

They kept digging. Peeling away the sediment layer by layer, they uncovered the partial skeleton of an individual who had stood no more than three feet six inches tall. It had arm and leg bones almost identical in size to Lucy's, but her species, *Australopithecus afarensis*, had lived in Africa more than 3 million years earlier.

Soon, more bones were found—the remains of an estimated eleven tiny individuals who died in this cave only 50,000 years ago. Researchers declared it to be a new species, naming it *Homo floresiensis*. The media called it "the Hobbit."

The paleoanthropological community was dumbfounded. Some rejected it outright, declaring the remains the result of disease or birth defects. Others suggested *Homo floresiensis* was a dwarfed version of *Homo erectus*.

Funny things happen on islands. In general, big things get small and small things get big. Flores was once home to two-foot-long rats, six-foot-tall storks, and elephants the size of ponies. Even today, it is home to the largest lizard in the world, the Komodo dragon. So perhaps Flores itself created these so-called Hobbits. Perhaps natural selection favored the survival of smaller individuals marooned on an island with limited resources. Perhaps inbreeding due to genetic isolation was a factor, turning an ancestral *Homo erectus* population into a relict species that lived until very recently.

But to others, these fossils suggest an even more extraordinary story.

Homo floresiensis's brain was smaller than that of *Homo erectus*. In fact, it was well within the range of an *Australopithecus*. It also stood at the same height and had limb proportions like an *Australopithecus*. It had a pelvis shaped like an *Australopithecus*. It shared foot and hand anatomies with *Australopithecus*. Perhaps the first hominins to expand outside of Africa were not members of our own genus *Homo*, but our predecessors.

When Chinese scholars digging at the 2.1-million-year-old stone tool site at Shangchen hit bone, perhaps they, too, will find small-brained, short-legged, large-footed hominins like the ones Tarus found in the Liang Bua Cave. Perhaps we have been too quick to believe that the long legs of the *Homo* genus were required to make the trek out of Africa. After all, short-legged *Australopithecus's* range stretched over 2,000 miles east from Chad to Ethiopia and almost 4,000 miles south from there to South Africa.

The walking distance from Ethiopia to the Cradle of Humankind caves in South Africa and the walking distance from Ethiopia to the Caucasus in Asia are about the same. Evolving long legs undoubtedly gave *Homo erectus* an energetic advantage as it expanded around the globe, but *Homo floresiensis* may be telling us that short-legged *Australopithecus* made the journey first.

If so, the Hobbits might not have been the only descendants of those first explorers.

In 2019, scientists working in a cave in Luzon, Philippines, discovered another tiny hominin that survived until relatively recent times. Only thirteen fossils—a few teeth, a femur, and a few foot and hand bones—have been discovered so far, but their shapes differ from anything previously known to science, including the Hobbits of Flores.

The discoverers named a new species: *Homo luzonensis*. It, too, was alive only about 50,000 years ago. However, on both Flores and Luzon, stone tools indicate that hominins had inhabited those islands for close to a million years. We can only imagine what the first *Homo sapiens* to arrive in the Philippines and in Indonesia would have thought of these tiny, small-brained, bipedal hominins.

As *Homo sapiens* were evolving in Africa, Neandertals were hunting game in Europe, Denisovans were fashioning tools on the mainland of Asia, and at least two species of small-bodied hominins inhabited the islands of Southeast Asia. The world, it seems, resembled nothing so much at Tolkien's Middle Earth.

Still, the story of human evolution was about to take yet another startling twist.

"WELL?" LEE BERGER asked as I emerged from a basement vault at the University of the Witwatersrand in January 2014.

I had been down there all day with more hominin fossils than I'd expected to see in a lifetime. There were no windows, and I had lost track of time. I hadn't eaten. My eyes were tired and swollen from staring at bones for hours. But I could still see the ear-to-ear grin stretched across Berger's face.

"I think it is a better *Homo habilis* than *Homo habilis*," I said.

Lee laughed. "Isn't it great?"

"Awesome."

That was the only word I could come up with, and it was not enough.

Five months earlier, I was working at my desk at Boston University when the all-too-familiar *ding* announced a new email. It was from Berger. The subject line said, "check this out," and a file was attached. I was pondering whether to break my rule of deleting emails with attached photos that invite me only to "check this out" when the phone rang.

"Jeremy, did you get it?" Berger asked. "What do you think?"

"Uh . . . just a second."

I fumbled to my mouse and clicked on the image of a partial skeleton—a mandible, some loose teeth, the side of a skull, a femur, a shoulder, and a handful of arm and leg bone shafts. They weren't embedded in rock or buried in dirt. They were just lying on the floor of a cave. That is how Hollywood thinks we find fossils, not how we normally find them.

"What do you think?" Berger asked.

"Just a second."

I needed to buy some time. My first thought was that it might be a dead spelunker. Should we call the police? But, no. Look at those teeth! No human has a wisdom tooth that large. Only early hominins had teeth like that.

"Jesus, Lee."

"Right?" He laughed his hearty Lee Berger laugh.

He was in a rush to share the news with other colleagues, so our conversation was short. I stared at my computer screen and it sank in: 7,853 miles away, a partial hominin skeleton was lying vulnerable and exposed in a cave.

On September 13, 2013, amateur cavers Rick Hunter and Steve Tucker were exploring the Rising Star cave system. Although it lies within a mile of the famous hominin fossil caves of Swartkrans and Sterkfontein in the Cradle of Humankind, no hominin fossils had ever been found there. Hunter and Tucker squeezed through a narrow crack, down a vertical chute, and into a chamber.

They saw bones everywhere.

When word reached Berger, he began planning an expedition to recover the fossils. This time, the work would require individuals with an unusual set of attributes. They would need excavation experience, caving know-how, and knowledge of comparative anatomy, of course, but they would also have to be slim enough to squeeze through the passages of Rising Star Cave, which at their narrowest are just over seven inches wide. Count me out.

Berger's solution was to send this message to the scientific community on Facebook:

We need perhaps three or four individuals with excellent archaeological/paleontological and excavation skills for a short-term project that may kick off as early as Nov. 1, 2013 and last the month if all logistics go as planned. The catch is this—the person must be skinny and preferably small. They must not be claustrophobic, they must be fit, and they must have some caving experience; climbing experience would be a bonus.

The post went viral, and Berger quickly found his team. Six women—Marina Elliott, Elen Feuerriegel, Alia Gurtov, Lindsay Hunter, Hannah Morris, and Becca Peixotto—were selected. Dubbed the "underground astronauts," they were asked to retrieve what was thought to be a partial hominin skeleton from the cave chamber. What they found was much, much more.

The all-female team retrieved over 1,500 hominin fossils from over a dozen different individuals—the largest discovery of hominin fossils from any site in Africa ever. Two months later, I was in Johannesburg to help figure out what had been pulled up from the depths of that cave.

The skulls were small, with brain sizes comparable to *Homo habilis*. They also had relatively small teeth, but like *Australopithecus* and early *Homo*, the wisdom tooth was the largest of the molars. The shoulders were shrugged like Lucy's, but the arms were shorter. The hand bones were quite humanlike except that they had curved fingers. The pelvis and hips looked like Lucy's. The legs were long like *Homo*, but they had relatively small joints. The feet looked a lot like yours and mine except that they were flat by modern human standards and had toes that were curved.

All told, it looked more human than *Australopithecus* but less human than *Homo erectus*.

This made it a candidate to be in the human lineage—the link between *Australopithecus* and *Homo erectus*. But was it? *Homo habilis* had already staked its claim for that role. With this combination of anatomies, I expected the new fossils to be around 2 million years old.

But there was something bothersome about the bones. Fossils can be as heavy as rocks, but these bones were light in weight. A few other fossils from South African caves felt light to the touch because they had been naturally decalcified by acidic groundwater. I assumed the same had happened at Rising Star.

After a year of study by an international team of forty-seven scientists, we announced to the world in September 2015 that these fossils represented a brand-new species of our genus. We named it *Homo naledi*.

Another year passed before our team of geologists figured out when *Homo naledi* had lived. The team used two approaches. First, the rate of radioactive decay in the surrounding limestone was tested to determine how long ago the bones had fallen into the chamber. In addition, bits of enamel taken from *Homo naledi*'s teeth were dated using electron spin resonance. This technique counts the electrons that have been bumped by radioactive particles and trapped in the crystalline structure. The longer something has been buried, the greater the number of trapped electrons.

The results of the two tests were consistent and shocking.

The bones are only 260,000 years old. In other words, *Homo naledi* lived at the same time as early members of our own species. That was why the fossils felt so light to the touch. They hadn't been in the caves long enough to have turned to rock.

Just a blink of an eye ago in geologic time, early humans shared the planet with *Homo naledi*, Neandertals, Denisovans, and the island Hobbits. And there is no question that they met, and in some cases interbred.

All of them walked on two legs, of course, but they walked a little differently from each other.

With short legs and long feet, *Homo floresiensis* walked like someone wearing snowshoes—high knees and short steps. They had to lift their legs to avoid tripping over their feet and may have had a hard time running.

We don't know much about *Homo luzonensis*, but a single fossil foot bone indicates that they had more mobility in the middle portion of their foot than our species. This would have compromised their ability to push off the ground, causing them to move as if they were wearing floppy slippers. However, it would have made them better climbers than us if or when they went into the trees for food or safety.

We know next to nothing about how Denisovans walked because we don't have enough bones yet. But Neandertals are another story. Their legs and feet were almost identical to ours, but subtle differences indicate that they were well suited for short bursts of speed and side-to-side movements over rough terrain.

And *Homo naledi*? Their bones suggest that they walked a lot like humans, but because they had flat feet and lacked large joints to dissipate the forces of impact, they wouldn't have had the endurance we have. As a result, they would have had small home ranges.

As recently as 50,000 years ago, different species of hominins walked the Earth, using their landscapes in slightly different ways. But the age of Middle Earth did not last.

Soon, there was only us.

We don't know why we are now the lone upright walking hominin. We do know that we didn't eliminate the Neandertals and Denisovans. We made babies with them and absorbed them into our gene pool. But the fates of *Homo naledi* and the island Hobbits remain a mystery.

PART III

Walk of Life

HOW UPRIGHT WALKING HAS SHAPED WHO
WE ARE FROM OUR FIRST STEPS TO OUR LAST

Afoot and light-hearted I take to the open road,
Healthy, free, the world before me,
The long brown path before me leading wherever I choose.

—WALT WHITMAN, "SONG OF THE OPEN ROAD," 1860

CHAPTER 10

Baby Steps

A journey of a thousand miles begins with a single step.

—*Lao Tzu,* Tao Te Ching, *sixth century BC*

In the middle years of the nineteenth century, French artist Jean-François Millet created several black-chalk-and-pastel drawings of a child learning to walk. He called them *Les Premiers Pas*, or "First Steps." Later, in 1889, the Dutch master Vincent van Gogh, who had checked himself into an asylum in Saint-Rémy, France, carefully drew even grid lines on a photograph of one of the drawings and began painting his own version on a fresh canvas.

Wavelike grass and trees with thick, squiggly leaves give the painting van Gogh's recognizable dreamscape. The farmer is in blue, except for his brown hat and shoes. His spade has been haphazardly tossed to his right; a wheelbarrow of hay sits to his left. His eyes are not visible, but he is clearly looking at his daughter. His hands are outstretched, arms fully extended, and I hear him say, "Walk to papa." The farmer's wife is also in blue. She bends at the waist, supporting her daughter as the little one leans forward to take those precious first steps. The girl grins mischievously, a glint in her eye. I imagine her giggling as she takes those steps.

When Vincent finished the painting in January 1890, he sent it to his brother Theo, whose wife Johanna was expecting their first child. It was a thoughtful gift from a troubled genius who would be dead by suicide six months later.

This painting, currently in the Metropolitan Museum of Art in New York, sings to us because it captures a moment that unfolds every day in every culture around the world—and has for millennia. The ubiquity of the scene does nothing to diminish the joy it gives and how momentous the occasion is for caregivers.

But how do children learn to walk, and why does it take so long for our little ones to figure it out?

After a long gestation, a baby is ready to be born. The female labors—sometimes for days—surrounded and supported by female relatives. The birthing female squats or kneels, using gravity to assist with the difficult birth. This could describe a human birth, but read on. Within an hour of delivery, the baby straightens its legs and takes its first wobbly steps. By the end of the first day, it can run to keep up with its mom and the rest of the group. It nestles in its mother's trunk and drinks her milk. The elephant herd moves on, plus one.

Many mammals, elephants included, give birth to young who begin moving through their environment almost immediately after birth. Baby seals and dolphins emerge from the womb already swimming. Baby giraffes and antelopes stand, walk, and run within twenty-four hours. This is necessary for their survival; many predators are on the hunt.

Other animals, however, are helpless at birth. Black bear cubs are no larger than your thumb. Nearly hairless and eyes still closed, they slowly crawl to their mothers' nipples to nurse and grow in the safety of the den until spring arrives. Many birds, too, give birth to helpless young who remain nest-bound for weeks.

Most primates, apes in particular, fall between the extremes of elephants and bears. They are born with fur, with their eyes open, and with some locomotor abilities. They can cling to their mothers soon after birth, but they rarely stray from them.

But humans are different.

For the first few weeks, human babies are luggage. They cannot walk like an elephant baby or cling like a chimpanzee baby, but they are not born as undeveloped as bears or birds, either. Right away, with eyes open, newborn babies are aware of their surroundings. They are drawn to familiar sounds, can mimic some facial expressions, and can socially manipulate an entire room. The long gap between birth and the onset of independent locomotion, however, requires a buffer against threats in the first years of life—something our ancestors would have needed, too.

Although human newborns cannot walk on their own, they practice the motions.

In May 2017, a video taken moments after a birth in Santa Cruz Hospital in Brazil went viral. It appeared to show a newborn girl walking. Her torso was slung over the arm of a nurse, her legs stretched downward, and her feet touched the tabletop. She lifted her left leg and took a step. Then she did the same with her right. Two motions: lift and step, left then right. Sure, she had support, but she was essentially walking despite being born minutes earlier.

"Merciful father. I was trying to wash her here and she keeps getting up to walk," a nurse is recorded saying in Portuguese. "Heavens above. If you told people what has just happened, no one would believe it unless they saw it with their own eyes."

The video, seen by 80 million people within forty-eight hours of its posting, is cute but unremarkable. It is not too unusual for newborns to go through the motions of walking. Albrecht Peiper, a German pediatrician, filmed babies in the first six weeks of life alternating their legs in what he called "primary walking." Other researchers have called this "upright kicking," "supine steps," or a "step reflex."

It does indeed appear to be a reflex, deeply ingrained in the mammalian body plan.

Seven to eight weeks after conception, a fetus begins to kick in utero. Alessandra Piontelli, who studies fetal development using ultrasound at the University of Milan, calls this "walking in the womb," and other scholars have proposed that this is energetically more efficient for the fetus than kicking both legs at the same time against the strong wall of the uterus. But do these movements have anything to do with walking?

At first, Nadia Dominici, a neuroscientist at Vrije Universiteit Amsterdam, did not think so. She figured that these steps in the womb and shortly after birth were eventually overwritten by a new, more sophisticated plan in walking toddlers. Surprisingly, however, her studies of how neuromuscular circuitry develops show that these first steps are foundational. They are a prototype that is refined and eventually perfected as a child learns to walk months later.

Think of the step reflex as programming a computer with two commands: extend the legs and alternate them left and right. Dominici's work reported that these commands are found not only in human neural circuitry but in other mammals, including rats. Apparently, alternating our legs is an ancient trait that we share with all our mammalian cousins.

If this step reflex in newborns lays the foundation for walking in toddlers, can strengthening the former impact the latter? About fifty years ago, Philip Roman Zelazo, a psychologist at McGill University, and his colleagues studied twenty-four newborn babies to find out.

For the first eight weeks of life, eight of these babies went through a daily exercise to practice and strengthen the step reflex, their parents holding them over a flat surface as they alternated their little chubby legs. The other sixteen babies did not. On average, those who practiced the step reflex took their first real steps at around ten months old, two months earlier than the others. Zelazo concluded that childrearing was more important to walking onset than a child's innate ability. Though this was a small study, Zelazo was onto something.

BABIES DO NOT come with manuals, but parents want to know if their child is developing on schedule. We seek advice from friends and family who have raised kids. I spent many nights flipping through a worn, dog-eared copy of Sears and Sears's *The Baby Book* passed down by my sister. But most new parents today do a Google search when they have a question. Googling "baby first steps" takes one to the U.S. Centers for Disease Control and Prevention website, where visitors are invited to "click on the age of your child to see the milestones." There, they can learn that by twelve months, their baby may take their first steps. The World Health Organization also reports that the average age by which a child walks independently is twelve months. But what if a baby is walking at nine months or hasn't taken a step yet at sixteen months? Is there something wrong? In most cases, no.

The average American child may take her first steps at around one year, but lost in the fine print is that the normal range is eight to eighteen months. If half of healthy children walk by their first birthdays, that means the other half do not.

Meanwhile, that one-year walking milestone has shifted over the years. It also differs from culture to culture.

Arnold Gesell, a pediatrician and psychologist at Yale University in the early twentieth century, was a pioneer in the study of child development. Although he argued that each child develops at their own pace, he still championed the idea of developmental milestones.

After collecting a lot of data, he found that in the 1920s, the average American child took her first steps between thirteen and fifteen months.

By the 1950s and 1960s, such developmental milestones became part of screening tests commonly performed in the pediatrician's office. Among them were the Bayley Scales of Infant and Toddler Development and the Denver Developmental Screening Test. These tools helped pediatricians identify developmental problems in a child, but two unhelpful things happened. First, many parents mistook "average" for "normal." Second, "earlier" was mistaken for "better." Parents actively encouraging their children to walk earlier resulted in a drop in the average age an American child walked to twelve months.

Things shifted again in 1992 when the "back to sleep" campaign was rolled out to counter the rising number of SIDS (sudden infant death syndrome) deaths. Researchers found that babies sleeping on their tummies were at a greater risk of dying from SIDS, so pediatricians recommended that babies be placed on their backs when they sleep. However, babies lying on their tummies develop stronger core musculature because of the way they adjust their bodies during sleep. As a result, they might be quicker to stand and take their first steps. A slight delay in standing and walking is a small price to pay for a reduction in SIDS deaths. Still, "tummy time" to strengthen a baby's core is now a recommended part of every day.

Clearly, when a child first stands and walks can vary depending on a number of factors. Even so, studies to determine the normal range have been done almost entirely on "WEIRD" populations: Western, Educated, Industrialized, Rich, and Democratic. As anthropologists Kate Clancy and Jenny Davis put it, "WEIRD is white."

Those studies have mistakenly been used to establish a baseline for what is "normal." When we examine the onset of walking through-out the world, even more variation can be found.

THE ACHÉ, A nomadic people from the forests of eastern Para-guay, rely on traditional hunting and gathering for food. They live in groups of about fifty people and eat what they can find, including palm starch, honey, monkeys, armadillos, and tapirs.

Their forest is dangerous, especially for children. Jaguars stealth-ily roam the forest floor. Venomous reptiles, including coral snakes, pit vipers, cobras, and the feared fer-de-lance, abound. In *Aché Life History*, anthropologists Kim Hill and A. Magdalena Hurtado write of biting ants, fleas, gnats, ticks, spiders, and even caterpil-lars. Wasp stings in this forest can cause vomiting, and a species of beetle produces an acidic liquid that burns the skin and can cause temporary blindness. Botflies deposit their larvae under human skin, Hill and Hurtado wrote, forming "an ever-growing and pain-ful wound which can contain a worm of alarming proportions." Be-yond the naked eye are more dangers: malaria, Chagas disease, and leishmaniasis—parasitic diseases caused by bites from mosquitoes, kissing bugs, and sandflies, respectively.

"An infant or small child would not survive long if left unat-tended on the forest floor," Hill and Hurtado wrote. "Forest camps are constantly interrupted by the cry of some child who is learning the hard way about which insects to avoid."

It would be dangerous for a parent to let a year-old Aché child learn to walk in this environment. So they don't. The children re-main tightly bound to their mothers for up to two years. Anthropol-ogists Hillard Kaplan and Heather Dove reported that the average Aché child is about two years old before she begins walking on her own. That's double the average age in the United States today. The

difference is cultural, not strictly biological. If my kids were being raised in the Aché's forest, they probably wouldn't have started
walking until they were two years old.

In parts of northern China, children are harnessed in bean bag–
like sacks of fine sand for sixteen to twenty hours a day so that they
can be left alone while adults tend the crops. At thirteen months of
age, an age when three-quarters of American children are walking,
only 13 percent of these Chinese children have taken their first steps.
In parts of Tajikistan, cradling of children prevents them from moving their legs much, and it is unheard of for one of them to walk on
their own at one year of age.

While there are plenty of cultures in which kids take their first
steps later than they do in America, there are others where they walk
sooner. In parts of Kenya and Uganda, for instance, it is not unusual
for babies to walk on their own at nine months. For years, such
differences were invoked to support the racist idea that Africans are
biologically different (read: inferior) to people of European descent.
Researchers know the reason these kids walk earlier has nothing to
do with their genes. Their mothers and grandmothers vigorously
massage the infants' legs during daily baths, and this stimulation
improves motor strength and coordination. A similar practice in
some neighborhoods in Jamaica has led to a similar result. The average kid there walks at ten months.

Still, even today a Google search reveals a swarm of misinforma-
tion about walking onset. "Late walkers are naturally smarter," boasts
one website. "The longer babies crawl, the smarter they are," claims
another. A third wonders the opposite: "Are children who walk and
talk early geniuses in the making?"

Developmental psychologists have studied this question, and the
results are ambiguous. A Swiss study of 220 kids found that those
who walked earlier had slightly better balance as eighteen-year-olds,

but they scored no better or worse on IQ tests or on tests of their motor skills. A larger, long-term study of over 5,000 people in the U.K. completed in 2007 found no relationship between when children took their first steps and their IQs at ages eight, twenty-three, and fifty-three.

Every once in a while, though, a study shows that children with a slightly higher IQ walked earlier. One problem with such research is that it is unclear what IQ actually measures other than one's skill at taking IQ tests. Besides, the effect, if real at all, is so small that it doesn't appear that the onset of walking has much to do with intelligence. If anything, the arrow of causality is in the other direction. Some researchers have posited that walking itself presents a toddler with a new view of the world and opens the door to new learning opportunities.

In 2015, however, researchers studying more than 2,000 kids in the U.K. found that those who were more active at eighteen months of age had denser bone in their shins and hip joints almost two decades later. Physical activity promotes bone growth. This could explain why a study of more than 9,000 kids in Finland found that children who walked at an earlier age were more likely to play sports as teenagers.

Still, the relationship between walking onset and athletic ability is weak at best and cannot be used to make predictions.

Muhammad Ali, when he was baby Cassius Clay, Jr., was already up, walking, and perhaps even throwing jabs at ten months of age. The greatest center fielder of all time, Willie Mays, took his first steps when he was a year old. Leroy Keyes, former professional football star and member of the College Football Hall of Fame, didn't start walking until he was three. Kalin Bennett, who was diagnosed with autism before he was a year old, also didn't take his first steps until he was three. But he took to basketball as a third grader, and by the time he was a high school senior, he was the sixteenth-best basketball prospect in Arkansas. He now plays college ball for Kent State University.

Three years is extreme, and it is strongly recommended that a child not walking independently by eighteen months be seen by a physician. But the point remains: the age of walking onset, within the expected range of eight to eighteen months, doesn't matter much at all.

TREMENDOUS VARIATION CAN be found not only in *when* kids learn how to walk but also in *how* they do it. The old adage that "you have to crawl before you can walk" isn't true at all.

Many children in cultures all over the world never pass through a crawling stage, and skipping it has no impact on their ability to learn to walk. A study of Jamaican infants found that nearly 30 percent of them never crawl. In England, one in five never do. Forty percent of early-twentieth-century middle-class infants in the United States never crawled because most were clothed in long gowns that caught on their knees if they tried, causing them to flop face-first to the floor.

Infants who do crawl don't do it the same way. They bear crawl, crab crawl, army crawl, spider crawl, belly crawl, knee-walk, inchworm, log roll, or bum shuffle. Eventually, they take their first steps.

"Individual infants forge their own paths," NYU developmental psychologist Karen Adolph wrote, "and the sequence of expression is variable." In other words, there is no one right way to become a biped.

To better understand how and why kids stand up and learn to walk, I visited Dr. Adolph's NYU lab in Greenwich Village. Adolph was a preschool teacher for six years before beginning her graduate work in developmental psychology at Emory University. Since then, she has been awarded more than forty grants to study early childhood development and written more than a hundred scientific papers, which have been cited by her colleagues more than 9,000 times. No one knows more about how kids learn to walk.

"Moving on all fours is fine for other animals and seems to be adequate for infants. Why, then," I asked, "do kids stand and walk on two legs?"

Adolph smiled and stared at me with piercing blue eyes.

"Why walk?" she said. "Why not?"

Ample data collected by Adolph's lab show that moving on two legs allows babies to go farther and faster. By equipping toddlers with cameras that capture the world from their point of view, Adolph also demonstrated that moving on two legs allows babies to see more of their surroundings. As Antonia Malchik wrote in her book *A Walking Life*, "Babies and toddlers are motivated to walk when there's somewhere interesting to go." Adolph's team has also found that walking babies carry objects forty-three times per hour—about seven times more than crawling babies.

"That makes sense," I added. "Walking frees the hands to carry objects."

"But that is not *why* they do it," Adolph quickly corrected me. "Our data show that they are not goal-directed."

She explained that babies aimlessly wander around the room, wasting all sorts of energy along the way. Do they eventually reach toys and other interesting destinations? Sure. But they take their time getting there.

"Why?" I asked.

"Babies move for the joy of it," she said.

I thought of my son Ben walking for the first time. (Thankfully, we have some of his first steps on video; otherwise, there is little chance my sleep-deprived brain would have recalled the moment.)

It was a warm August afternoon, and my wife and I were trying to stay cool in our little house in Worcester, Massachusetts. Our twins had been crawling for months and could pull themselves up to stand and shuffle along the couch or a bookshelf. My son seemed intent on walking. He would extend his legs and take an uneasy step or two before his chubby thighs collapsed his knees and he whomped onto his rear. His twin sister watched with delight, but she rarely tried to walk herself.

Scientists who study how kids grow up define first steps as taking five unaided steps without falling.

Ben was in his dark blue Red Sox onesie, his large bald head perched precariously on his wobbly body like a baby version of Charlie Brown. My wife held his hands above his head and then gently pulled away as he began to stumble toward my outstretched arms. With each lift of a leg, he drew closer, and the closer he got, the harder he laughed. After five steps, he wore a huge smile as he collapsed, giddy and exhausted, into my arms.

Yes, babies move for the joy of it.

Of course, after that moment, Ben didn't walk everywhere he went. He continued to crawl, scoot, and cruise. Walking was just another tool in his locomotor tool kit, but before long, it was the dominant one. My daughter Josie watched closely, and, not to be outdone, she soon joined him. Being able to watch and mimic others might have a lot to do with how kids learn to walk. It might also explain why seeing-impaired children take twice as long on average to take their first steps.

It is said that to become an expert at anything—mastering a musical instrument, or playing a sport—one has to invest 10,000 hours. Learning how to walk is not that different.

"How do you learn to walk?" Adolph wrote. "Thousands of steps and dozens of falls per day."

The oldest known bipeds. From right to left: *Cabarzia*, *Eudibamus*, and *Lacertulus* from the early Permian (around 290 million years ago). *Courtesy of Frederik Spindler.*

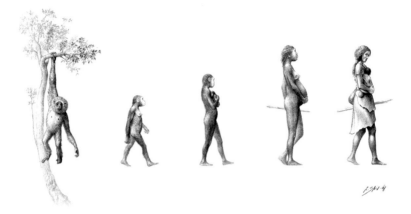

A simplified view of bipedal evolution. *Courtesy of Eduardo Saiz Alonso.*

Artistic reconstruction of *Danuvius guggenmosi*, an 11.62-million-year-old upright ape from the Hammerschmiede clay pit in Germany. *Courtesy of Velizar Simeonovski.*

Early hominin bipedal footprint from the 3.66-million-year-old site at Laetoli, Tanzania. *Image taken by author.*

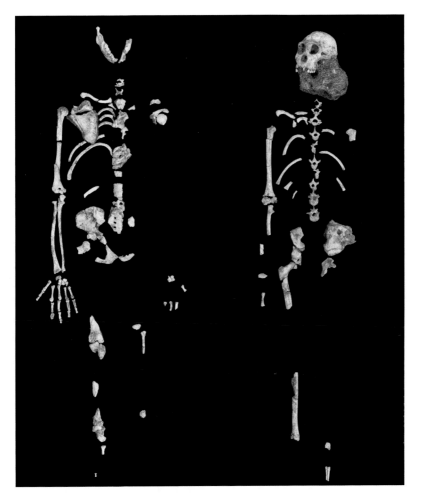

Nearly 2-million-year-old skeletons of *Australopithecus sediba* from Malapa Cave, South Africa. *Courtesy of Getty Images/Brett Eloff.*

Vincent van Gogh's *First Steps, after Millet. Metropolitan Museum of Art, New York.*

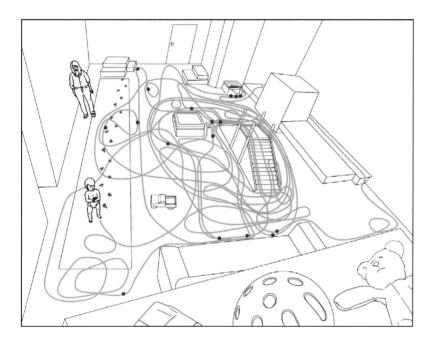

The trail left by a thirteen-month-old toddler spontaneously walking for ten minutes in Dr. Karen Adolph's developmental psychology lab at New York University. The dark blue dots indicate moments the child stopped. *Courtesy of Karen Adolph.*

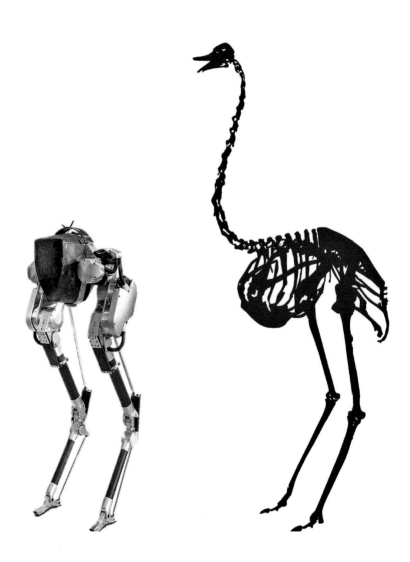

Cassie —a bipedal robot designed by engineering professor Jonathan Hurst—compared to the skeleton of an ostrich. *Image of Cassie courtesy of Jonathan Hurst and Mitch Bernards. Image of ostrich courtesy of Getty Images/ iStockphoto.*

Early in her career, Adolph observed that kids do not walk in straight lines, yet laboratory equipment, such as treadmills and gait carpets used to measure walking, were always straight. To get the data she wanted, she used her entire lab space, recording every step new walkers make. What she and her student assistants found is remarkable.

The average toddler takes 2,368 steps per hour, covering nearly the length of eight football fields. In a typical day, then, toddlers take about 14,000 steps—enough to score forty-six touchdowns or cover close to three miles. No wonder they need at least twelve hours of sleep daily.

Toddlers don't walk like little adults. They take uneven steps, wobbling from side to side with slightly bent hips and bent knees like miniature apes. Their feet are flat, and they don't push off the ground efficiently. Adolph's team found that they average seventeen falls an hour, but the thousands of daily steps they take help them improve. Even so, they don't begin to walk like adults until age five to seven. In the process, their skeletons change.

BONE IS ALIVE.

In a science classroom, bone specimens are hard, brittle, and un-yielding. They typically have an off-white color. But in you, bone is more pliable and dynamic. It is made in part of living cells that breathe and rely on hormones to receive messages from other parts of your body. Living bone has a blood supply that makes it very light pink in color.

Intuitively, you know that bone is alive. It grew in you when you were a tiny baby and eventually formed the skeleton you have now. If you break a bone, you know it can repair itself.

You have the same number and kind of bones as a chimpanzee.

That number is typically 206, though it can vary a bit depending on how many accessory bones form. Kids, however, have more "bones" than adults. Take the femur, for instance. In an adult, it is one bone—the biggest bone in the body. But in kids, it consists of a shaft and four bony knobs—three at the hip and one at the knee end. They are separated from the shaft by growth plates, which are regions of cartilage that proliferate as an individual grows. All of this is true for the African apes, too.

So what makes our skeleton suitable for upright walking while a chimpanzee's is not?

Genes are part of the story. The genetic code for where and how much cartilaginous scaffolding will grow in a developing fetus, whether human or chimpanzee, helps determine the architecture of the skeleton in a newborn baby. Certain anatomies of our skeleton make us, in some ways, born ready to walk.

For example, newborn human babies have chunky heels already prepared for the rigors of upright walking. From birth, our pelvis is short, stout, and positioned on the sides of our body, anchoring the muscles around the hip joints that help us balance when walking on two legs. Newborns even have the spidery network of spongy bone on the inside of the pelvis aligned to transmit the forces of upright walking. This is not something babies *need* for another year, but they are born with it. These, therefore, are true genetic adaptations for walking on two legs.

But remember, bone is alive. Its cells respond to forces imposed upon them as you grow. In a way, they remember all of the times you have moved and *how* you move. As children develop, their bones do not just increase in size. They change shape in response to the daily stresses kids put them through.

Take the knee, for example.

When Ben took his first steps, he wobbled from side to side in part because his knees remained quite far apart. But in older children and adult humans, the knees practically touch, helping balance our bodies by keeping our feet under our hips. This happens because our femurs are angled inward. This came up in Chapter 4 since Lucy and her kind had this bicondylar angle in their femurs, too, but this is not something we are born with, and neither was Lucy. At birth, our femurs are pin-straight, like a chimpanzee's. As we begin to walk as toddlers, the cartilage of our knees receives uneven pressure and grows at an angle, resulting in a tilted knee. Individuals who are paraplegic and have never taken a step never develop this angle.

However, there are always trade-offs in evolution. The bicondylar angle, while beneficial for keeping the body balanced, can cause problems. Because our femur is angled, the quadriceps muscles anchored to the front of this bone contract at an angle when we move. The result is a sideways force that pulls the kneecap slightly out to the side. In extreme cases, it can dislocate the kneecap—something doctors call "patellar subluxation."

Given the physics involved, it might seem that kneecap dislocations should happen more frequently than they do (around 20,000 a year in the United States). They don't because a large ridge of bone called a lateral patellar lip acts as a retaining wall, keeping the kneecap in place. This is the same anatomy that was unusually large in the knee of *Australopithecus sediba*. You can feel this on yourself if you sit with your knees bent and rub the outside top of your knee.

The remarkable thing about the lateral patellar lip is that like the spongy bone on the inside of the pelvis, we are *born* with it even though we don't need it until we start walking. The cartilaginous scaffolding for the lateral patellar lip is already in baby knees at birth—a solution to a problem infants don't have yet.

This is a wonderful example of how our bodies are a combination of genetically encoded traits and anatomies molded by our own be-havior. Our skeletons are the product of nature and nurture work-ing together in a complicated dance that results in the human form.

BACK IN HER lab, Adolph showed me videos of kids crawling along an elevated trackway. Their eyes were fixed on a stuffed an-imal being used to lure them along. What they didn't notice was the footwide gap in their path. If not for a spotter, they would have plunged right off the trackway. Adolph and her team varied the chal-lenges for the crawlers, adding slopes or other obstacles. The results were always the same. Infants exploring their world for the first time are fearless and have no sense of their limitations. Soon, however, they learn from their mistakes and begin to crawl with more cau-tion and awareness of the pitfalls around them.

That is, until they start walking. I watched in amazement as video after video showed the very same kids navigate obstacles just fine on all fours but foolishly wobble right off the experimental walkway when they were bipedal.

"Wow—they forgot everything they learned," I said.

"No," Adolph replied. "They learned how to navigate these ob-stacles while crawling, but walking gives them a new vantage point on the world. Our view of the world is contingent on how we move. They only know what they know in the context of a particular type of locomotion."

As I watched videos of child after child falling into gaps in the walkway, stepping over the brink of high drop-offs, and plunging down steep slopes, I was grateful for the ever-present hands of spot-ters catching these fearless new bipeds.

Learning to walk is hard and even dangerous—unless someone is there to catch you.

CHAPTER 11

Birth and Bipedalism

these hips are mighty hips
these hips are magic hips

—*Lucille Clifton*, Homage to My Hips, *1980*

She stood upright. Labor pains made her fists clench, and her arm muscles tensed, becoming well defined under her skin. Occasionally, she squatted or sat while two other women supported her. One was positioned behind her, arms wrapped under her breasts, helping hold her up while she breathed and braced for the next wave of uterine contractions. They whispered encouraging words to her and told her that the baby was nearly out. They knew because they had done this before.

She pushed one last time, and her baby, still attached to her via the umbilical cord, entered a cold, dangerous world. Her sisters wrapped the infant, and the exhausted mother brought the baby to her chest to nurse.

These events could have happened today, but what is described above occurred just over 15,000 years ago. We know this because someone carved the scene—the oldest depiction of childbirth in the archaeological record—into a piece of slate and left it with other ancient carvings of daily life in what is today Gönnersdorf, Germany, a town south of Bonn, just a few miles west of the Rhine River.

Fifteen thousand years ago, our planet was beginning to emerge from the most recent ice age. Scandinavia and the British Isles were still covered with Arctic ice a mile thick. Today, undeveloped regions of Germany are forested, but at the end of the Pleistocene, there was not a tree for miles. It was tundra, similar to today's Siberia. The area was inhabited by animals found in polar grasslands today, including caribou, Arctic foxes, and musk oxen. There were also now-extinct animals, including mammoths, woolly rhinoceroses, and cave lions. The people made tools and fire. They hunted and cooked. They had babies and made art.

These last two activities melded into one marvelous event that yielded the carved piece of slate we call Plaquette 59. It was important to the carver to capture the laboring woman's contracting deltoid muscles and to show her fingers squeezing together. But there is abstraction here as well. The newborn is depicted as a simple oval with eyes connected to her mother by a squiggly line. Two horse heads, the meaning of which died along with this culture, look on.

Childbirth has always been part of human life, and it is intimately, but complicatedly, linked to upright walking—specifically how women walk.

EVERY WOMAN WHO gives birth has a unique experience. Every birth is a complicated mixture of fetal head size and shoulder width, pelvic dimensions, gestation length, ligament relaxation, cranial molding, stress hormones, birth position, social support, and the advice and approach of the attending midwife or obstetrician, among other variables. But in this sea of variation there are commonalities, especially when human birth is compared to how our closest living ape relatives do it.

In humans and the great apes alike, the baby is typically head down in the uterus and facing forward (toward the mother's belly) by the end of the third trimester. We don't know too much about birth in wild apes since females often deliver in trees, alone, and almost always at night, but such births have been closely observed in captivity.

Female apes have short labors, usually around two hours. The baby proceeds unimpeded through the bony birth canal and usually emerges facing forward. If positioned in the right way, a mother ape could look her own baby in the face as she pushes it from her vagina. Ape mothers can reach down and use their hands to help pull the baby from the birth canal. They lick and clean the newborn's face, helping clear airways. Nursing begins soon after.

Birth in humans is rarely so simple. A human baby typically starts as an ape does: head down and facing forward. On average, labor lasts fourteen hours, but it is not unheard of for a woman to labor for forty hours or more. Some of this time can be accounted for by the slow dilation of the cervix—the junction between the vagina and the uterus—required to pass a head with the dimensions of a human newborn.

When my mother was in labor with me and my head had reached the bony rim of her pelvis, I encountered my first obstacle. The pelvis tilts forward in our bodies so that the bony rim of the birth canal is angled. The pelvis's dimensions, from the top to the bottom, are usually too restricted for a human baby to be born the way other primates are born. The solution I, and just about every other baby, figured out was to tuck my chin to my chest and rotate my head to the side so that the longest dimension of my head (front to back) aligned with the widest dimension of my mother's pelvis (side to side).

In 1951, University of Pennsylvania anthropologist Wilton Krogman wrote an influential *Scientific American* article titled "The Scars of Human Evolution." In it, he argued that we can blame evolution for our many imperfections, from our sore backs to our crooked teeth. On birth, he wrote, "there can be no doubt that many of the obstetrical problems of Mrs. *H. sapiens* are due to the combination of a narrower pelvis and a bigger head in the species. How long it will take to balance that ratio we have no idea. It seems reasonable to assume that the human head will not materially shrink in size, so the adjustment will have to be in the pelvis; i.e., evolution should favor women with a broad, roomy pelvis."

But the problem is not that women have narrow pelvises. Side to side, they are sufficiently roomy. The problem is that the human pelvis is squashed from top to bottom, so the path by which primates typically are birthed doesn't work for us.

Why not?

Because we walk on two legs.

Apes have a tall pelvis, similar in shape to the pelvis in most other quadrupedal mammals. The hip joint is positioned far from the sacroiliac joint that connects the spine to the pelvis, resulting in a birth canal that can easily accommodate a baby's head. But this anatomy also makes apes top-heavy, wobbly, and unstable when they stand on two legs.

As our ancestors became more reliant on bipedal locomotion, their pelvis changed shape. In fact, the pelvis changed more than any other bone in our body, evolving from tall and flat to short and stout. A shorter distance between the sacroiliac joint and the hip joint lowered the center of mass in our bipedal ancestors, making them more stable and efficient upright walkers. But the shortening of the distance between our back and our hips reduced the size of the birth canal. Once this happened, babies had to turn their heads to the side and begin to rotate during delivery.

We can tell from the shape of Lucy's pelvis that this mechanism of birth goes back well over 3 million years.

On the morning of April 7, 1976, my mother's uterine contractions continued to push me through the birth canal and into an area called the midplane, where I encountered obstacle number two. Two bony projections called ischial spines make this part of the birth canal narrow from side to side. In fact, the birth canal went from being widest from side to side to being narrowest in that dimension. In most female pelvises, this is the narrowest point a human baby will encounter. The only way through is to keep rotating.

"Navigating the birth canal," anthropologist Karen Rosenberg has said, "is probably the most gymnastic maneuver most of us will ever make in life."

These changing dimensions of my mother's birth canal caused me to corkscrew through the midplane and into the pelvic outlet. By twisting my way through the birth canal, I was now facing my mother's back. In a squatting position, a mother could look down at this point and see the back of her baby's head crowning. This is called an "occiput anterior" birth, meaning the back of the head faces forward. Sometimes, however, a baby does not rotate as described above and instead is born facing forward with the back of the head pressed against the lower spine. This is called a "sunny-side-up" birth, and it happens about 5 percent of the time.

Occiput anterior births, the most common way humans are born, involve the fewest complications. But there is a trade-off. If my mother had reached down and tried to assist her occiput anterior baby out of the birth canal as ape mothers do, she would have risked pulling my neck backward and inflicting severe injuries.

At this point in my birth, I had crowned, but I was not born yet. I had to get my shoulders out. As Alice said when presented with a small door in Lewis Carroll's *Alice's Adventures in Wonderland*, "Even if my head would go through, it would be of very little use without my shoulders."

It is not uncommon for complications to occur at this point because the baby's wide shoulders, oriented perpendicular to the head, can get stuck in the bony pelvis. The trick is for the baby to birth one shoulder at a time by dipping its lead shoulder under the front part of the mother's pelvis. Again, a midwife or obstetrician can assist with this maneuver. After my shoulders were through, the rest of my body was born with ease, and my life began.

Because our hominin ancestors had a pelvis shaped a lot like ours, they, too, would have needed help giving birth. In fact, Rosenberg, now a professor at the University of Delaware, and Wenda Trevathan, a New Mexico State University anthropologist who has assisted at hundreds of births as a midwife, propose that rotational birth in hominins *required* helpers. For them, as with every human culture today, childbirth must have been a social event.

Yet human birth, even with the assistance of midwives or obstetricians, can still be dangerous.

"Childbirth is beautiful," Angela Garbes, author of *Like a Mother: A Feminist Journey Through the Science and Culture of Pregnancy*, wrote, "but it is not pretty. It is grisly and life affirming, glorious and deadly."

Worldwide, nearly 300,000 people and 1 million babies die annually in childbirth. For the mother, hemorrhage or infection are the primary causes. Countries where these death rates are the highest are usually the poorest and ones in which women have the fewest reproductive rights.

Maternal mortality is particularly high where child-bride practices are common and girls give birth before their bodies are done growing. According to a 2019 United Nations Human Rights Council report, it is the leading cause of death for fifteen- to nineteen-year-old girls in developing nations. In countries where the average marriage age for women is twenty years or more, the maternal mortality rate averages 1 in every 1,500 births. But in countries where the average marriage age is less than twenty, the average maternal mortality rate is an alarming 1 in every 200 live births. That's 7.5 times greater.

In the United States, about 700 women die in childbirth annually. That is about 1 in every 5,000 births. For a modern society, it is not an excellent number, making America the forty-sixth most dangerous country in the world for a woman to give birth—a bit better than Qatar and a bit worse than Uruguay. And it is getting worse.

Today, American women are 50 percent more likely to die in childbirth than their mothers were. Obstetricians say that's partly because reproductive health care has become harder to come by due to skyrocketing medical costs, difficulties acquiring affordable health insurance, and the closing of women's health clinics in disputes over abortion. Institutional racism at many points in the process makes women of color three to four times more likely to die in childbirth than white women. For every death, there are one hundred close calls in which emergency surgery and blood transfusions are required to save the mother's life.

Given these high mortality rates, one may wonder why evolution has not fixed this problem. The answer is complicated and unclear, but it starts with an idea known as the "obstetrical dilemma."

I AM PROBABLY the only anthropologist who knew about Brad Washburn before I knew about his brother, Sherwood Washburn.

Brad Washburn was a cartographer who mapped the White Mountains in New England and helped map Everest and other Himalayan peaks. His wife Barbara, no less an explorer than Brad, was the first woman to summit Denali (formerly Mt. McKinley) in Alaska. More important for me, though, was that Brad was the founder of the Boston Museum of Science. I worked there as a science educator from 1998 to 2003. It's where I met my wife, rediscovered my love for science, and found my passion for paleoanthropology.

One day over lunch in 2001, Brad and Barbara Washburn told me stories about the early days of the museum from Spook the owl to how the world's largest Van de Graaff generator ended up in the museum's parking lot. Then Barbara asked about my interests, and I launched into my newfound passion for human fossils.

"You know," Brad said, "my brother Sherry was an anthropologist."

I had no idea at the time, but Brad's brother Sherwood (Sherry) Washburn is a legend in the field. His Ph.D. advisor at Harvard, Earnest Hooton, spent a career identifying differences between human populations and clustering people into racial categories. But Sherry Washburn saw something very different in the data. He saw human variation as continuous and seamless, not categorical. This new approach to anthropology, explicated in his 1951 classic *The New Physical Anthropology*, changed our field forever and for the better.

Sherry Washburn also argued that studying living primates could teach us something about the behavior of our hominin ancestors. When molecular studies showed that humans were most closely related to chimpanzees, he endorsed a knuckle-walking pathway to bipedalism. He wrote about stone tools, fossil *Australopithecus*, and baboons. But mostly, he was interested in the behavior of early humans.

In 1960, Sherry Washburn wrote a piece for *Scientific American*. While the focus was ancient human technology and social behavior, his lines about human childbirth have gripped our field for sixty years. He wrote:

In man adaptation to bipedal locomotion decreased the size of the bony birth canal at the same time that the exigencies of tool use selected for larger brains. This obstetrical dilemma was solved by delivery of the fetus at a much earlier stage of development. But this was possible only because the mother, already bipedal and with hands free of locomotor necessities, could hold the helpless, immature infant.

A few sentences later, Washburn referred to the "slow-moving mother," unable to hunt with a baby in her arms.

Hence the phrase "obstetrical dilemma," which succinctly describes a classic evolutionary tug-of-war. The female pelvis has to be large enough to birth a newborn, but if too large, it would compromise locomotion. Evolution's solution was a pelvis just large enough for birth to happen, sometimes with difficulty, but not so large that a woman couldn't walk. To make things a bit easier, it was thought, babies are born early and smaller, but also more helpless.

In his influential book *Sapiens*, historian Yuval Noah Harari amplified Washburn's hypothesis. Bipedalism required a constricted birth canal, he wrote, "and this just when babies' heads were getting bigger and bigger. Death in childbirth became a major hazard for human females. Women who gave birth earlier, when the infant's brain and head were still relatively small and supple, fared better and lived to have more children."

Washburn's obstetrical dilemma is an elegant evolutionary hypothesis, but that doesn't make it right. Today, a new generation of researchers is challenging its assumptions.

To test the idea that humans are born early so they can fit through the constricted birth canal, University of Rhode Island anthropologist Holly Dunsworth and her colleagues compared the gestation lengths of various primate species. Gorillas gestate for about thirty-six weeks, chimpanzees and bonobos for between thirty-one and thirty-five weeks, and orangutans for between thirty-four and thirty-seven weeks. But humans typically gestate for between thirty-eight and forty weeks, which is over a month longer than one would expect for a primate of our size.

Humans do not give birth earlier than our fellow primates. They give birth *later*. During an elongated third trimester, the fetus packs on subcutaneous fat, its brain grows larger, and it demands more and more energy from its mother. In their 2012 study, Dunsworth and her colleagues hypothesized that birth is triggered when the energy demands of the growing baby exceed the metabolic capacity of the mother.

As birth grows near, an infant's brain is anything but small. Human newborns' brains average 370 cubic centimeters—the same size as an *adult* chimpanzee brain. Yes, our infants are born relatively helpless, but not because they are born early.

Why, then, have human females not evolved more spacious pelvises to make birth easier and safer? It would require only a couple of centimeters at each of the bony obstacles—a slightly taller bony rim and slightly wider ischial spines.

The long-held explanation assumed, without much evidence, that because women's bodies are adapted for childbirth, they are not as good as men at walking. That, it was thought, was the evolutionary trade-off for having large-brained, large-bodied infants. Enlarging the female pelvis, it was believed, would make walking untenable. Only recently has this idea been tested, and it, too, appears to be flawed.

"Initially, I totally accepted it," Anna Warrener, an anthropologist at the University of Colorado, Denver, said when I asked if she had always been skeptical of the assumption that women's gaits were compromised because of childbirth. But, she said, "no one had collected the data" to test the idea.

As a graduate student at Washington University in St. Louis, she teamed up with Herman Pontzer, now a professor at Duke University, to do just that. Warrener is not only an anthropologist but also a ballet dancer, a skill that makes her acutely aware of subtle differences in the way people move. She put women and men on treadmills and measured how much CO_2 they exhaled as they walked. If they exhaled more, they were using more energy. She also put her study participants through an MRI to measure their pelvises. According to Washburn's obstetrical dilemma, those with the widest hips use the most amount of energy.

But, they don't. In 2015, Warrener reported that the predicted relationship between hip width and energy efficiency just isn't there.

To understand what might be going on, I went to Seattle Pacific University on a cold, icy February morning to visit anthropologist Cara Wall-Scheffler. Two inches of snow had fallen in Seattle overnight, and the city was paralyzed. As a New Englander, I shrugged off the weather and walked to her office, which was overflowing with books, papers, replicas of hominin fossils, and her kid's LEGOs. A miniature plastic pelvis sat on her desk. Undergraduate researchers kept popping their heads in to update her on the walking experiments in the lab next door.

As a graduate student at the University of Cambridge, Wall-Scheffler had grown interested in Neandertals—in particular, a spectacular Neandertal skeleton from the Kebara Cave in Israel. The Kebara Neandertal died about 60,000 years ago and was deliberately buried by members of his group. The fragile partial skeleton includes the hyoid, ribs, and a nearly complete pelvis.

When Wall-Scheffler examined this pelvis, she was puzzled.

"Here was this pelvis, an enormous, wide pelvis in a male Neandertal," she told me. "No one was arguing that Neandertals were compromised in their walking abilities because of a wide pelvis. I was troubled by this and by the idea that women are compromised—

that women were worse walkers. I just thought that was wrong. Women are the evolutionary bottleneck. They are the unit of selection—women carrying kids. Why would evolution compromise their walking? Make them less efficient? It made no sense from an evolutionary standpoint."

Furthermore, research on hunting and gathering communities exposed the fallacy of Washburn's "slow-moving mother." Women from the Hadza of Tanzania to the Pumé of Venezuela cover an average of almost six miles a day. It wouldn't make sense for women who walk that much to have evolved anatomies that make locomotion inefficient.

In fact, researchers were finding evidence that natural selection has fine-tuned the female skeleton for the unique mammalian challenge of being pregnant bipeds.

In 2007, Katherine Whitcome and Daniel Lieberman of the Department of Human Evolutionary Biology at Harvard University and anthropologist Liza Shapiro of the University of Texas, Austin, studied what happens to a woman's gait and posture during pregnancy. As gestation reaches the third trimester, a sizable amount of baby, placenta, and amniotic fluid accumulates in the front of the body, pulling the center of mass forward. Quadrupedal mammals don't have the same problem since weight gain during pregnancy wouldn't alter the center of mass.

The Ig Nobel Committee, which honors research that may seem absurd at first but turns out to be important, made light of this, calling Whitcome's paper a study of "why pregnant women don't tip over." Actually, it is a great question: how do women adapt as their center of mass changes during pregnancy? It turns out that the answer lies in the small of the lower back.

Men and women both have five lumbar (lower back) vertebrae. In men, the bottom two are shaped like wedges, producing a curve in the spine that brings the torso over the hips. But, in women, the bottom *three* vertebrae are wedge-shaped, giving them a larger curve. Whitcome found that this helps pregnant women bring the shifting center of mass back over the hip joints, keeping them balanced as they walk.

This sex difference in the shape of the third-to-last lumbar vertebra occurred early in our evolutionary past. *Australopithecus*, Whitcome found, had it over 2 million years ago.

Meanwhile, Wall-Scheffler has consistently found that women walk as efficiently as men. But in some circumstances, she has discovered, they are even better at it.

As an evolutionary anthropologist, she didn't want to limit her studies to people on treadmills. We don't walk only in straight lines or on flat surfaces, and neither did our early ancestors. We don't always walk empty-handed, either, and neither did they. With hands freed by bipedal walking, they carried food, water, tools, and babies. When Wall-Scheffler measured how much energy we use when we are carrying, what she found fundamentally altered what we once thought about the female pelvis and the obstetrical dilemma.

Walking while carrying an object roughly the same size as a human infant, she found, can increase the amount of energy expended by nearly 20 percent. But the energy required is significantly reduced in individuals with wide hips from side to side—the kind often found in women.

"Women," Wall-Scheffler told me, "are, by all measures, better carriers than men."

In other words, wide hips are not about childbearing. They are about child *carrying*. But there is more.

When walking, humans can settle into their most efficient pace and walk great distances without using too much energy. But walking with a group, especially one that includes children, often means slowing down, stopping, and speeding back up. When men vary speeds, they use more energy, Wall-Scheffler found, but wider hips make it easier for women.

My wife used her hips as a shelf, perching our babies there while she walked around the house. When I tried it, the babies slid down my thigh. With no bony shelf, I had to carry my twins in my arms, and after a while, my arms tired. Carrying squirmy kids is not easy, never mind if you are walking six miles a day as modern hunters and gatherers do.

Wide hips are not a detriment to a woman's gait. They are adaptive. They also affect *how* many women walk.

Wall-Scheffler, Whitcome, and other researchers have found that having wider hips causes more rotation, or swivel, when walking. This allows women, who typically have shorter legs than men, to take longer strides than one might expect. Wider hips don't make a women's walk less efficient. They just make it mechanically different.

CLEARLY, LOCOMOTION IN women is not compromised by their wide hips, but maternal deaths during childbirth remains a problem that evolution has left unsolved. Why? We don't know, but researchers have posed several hypotheses that are in need of rigorous scientific testing.

One idea is that the high mortality rate might be a recent phenomenon. These days, many people subsist on a diet rich in simple sugars. This leads to big (macrosomatic) babies. It also may stunt the growth of girls, including their pelvises, during adolescence. Bigger babies and smaller pelvises is not a good mix.

Other researchers suggest that the problem may have something to do with the climate when and where our early ancestors evolved.

Today, people who have lived for generations in cold climates tend to be short and stocky and have wide hips because that body shape helps them stay warm. Closer to the equator, bodies tend to be narrower because that body shape helps them stay cool. Since *Homo sapiens* evolved in Africa, much of which lies near the equator, the earliest members of our lineage may have faced an obstetrical dilemma in which the need to stay cool limited the size of the birth canal.

Another hypothesis involves the anatomical relationship between the birth canal, the hips, and the knees. Pushing the hip joints farther apart could allow the birth canal to be wider, but it would also mean the angle of the femur would change in order to keep the knees directly under the torso for efficient upright walking. That would put so much pressure on the knees that the risk of debilitating anterior cruciate ligament tears would be intolerable.

A final hypothesis has been pitched by anthropologist Wenda Trevathan, who suggests that in trying to understand the relationship between upright walking and birth mechanics, we have focused too much on the walking part and not enough on the upright part.

Her idea involves pelvic organ prolapse, a potentially debilitating condition in which the uterus, bladder, or the lower part of the digestive system protrudes into the vagina. This occurs when the ligaments and muscles of the pelvic floor stretch during pregnancy and delivery and do not fully recover. Some studies suggest that pelvic floor muscles are torn in one in every three births, and that pelvic prolapse impacts 50 percent of women worldwide. This can happen in quadrupeds, too, but their internal organs rarely protrude because their birth canals are parallel to the ground, making them relatively unaffected by the tug of gravity. But the tug is significant in upright walkers.

Widening the space between the ischial spines, the narrowest part of a woman's birth canal, would ease the difficulty of giving birth, but it would also increase the risk of organ prolapse. Perhaps, then, narrow ischial spines were an evolutionary trade-off to strengthen the pelvic floor.

Maybe, as with Washburn's obstetrical dilemma, none of these hypotheses will withstand scientific testing. It's one of the hottest topics in biological anthropology today.

ALTHOUGH PREDICTIONS GENERATED by the obstetrical dilemma don't hold up under scrutiny, men obviously perform better than women in athletic events that involve running. Or do they? To explore this, we have to dive deeper.

In the early 1950s, running enthusiasts eagerly anticipated the eclipse of two barriers. Most famous was the four-minute mile, which was finally achieved by Roger Bannister in May 1954. The other was the two-hour twenty-minute marathon. The marathon world record had been stuck in the 2:20s for almost thirty years when Jim Peters, a thirty-three-year-old former Olympian from England, ran London's 1953 Polytechnic Marathon in two hours, eighteen minutes, and forty-point-two seconds (2:18:40.2). Today, Kenyan Eliud Kipchoge holds the official world record at 2:01:39. Now, a new goal is in sight: breaking the two-hour mark.

The year Peters set his marathon record, the women's record was 3:40:22, set by Violet Piercy on the same course in 1926. That record stood for almost forty years. Why? Because women were almost always banned from competing.

The Boston Marathon didn't see its first female entrant, Kathy Switzer, until 1967, and even then, a race official attempted to physically remove her from the course. The Polytechnic Marathon didn't have a women's division until 1976, and the women's marathon was not an Olympic event until 1984. Even so, elite women athletes lowered the world record by over an hour between 1964 and 1980. In those years, only three minutes were dropped from the men's marathon record.

Access and opportunity matter. Today, the women's marathon record is held by Brigid Kosgei of Kenya, who won the 2019 Chicago Marathon in 2:14:04. If we could push the bookends of time together and put her on the 1953 Polytechnic course with Jim Peters, he would be a mile behind when she crossed the finish line. In fact, Kosgei would have held the marathon record for both sexes until 1964.

To be sure, the top men continue to outcompete the top women in races from the hundred meters to the marathon because they tend to have more muscle mass and lung volume. In fact, the difference between the world records is almost always about 10 percent. But when we step away from the elite athletes and look at average folks, the differences between men and women turn out to be greatly exaggerated.

On a beautiful New England day in the autumn of 2012, I lined up with over a thousand other runners, hoping I could reach my personal goal of running a marathon in less than four hours. I crossed the line in three hours and fifty minutes, a respectable time slightly faster than the average for all competitors that day. Finishing ahead of me were 128 women, roughly 30 percent of female entrants. Sure, the overall top male was faster than the overall top female, but among everyday Joes and Janes, the difference in athletic ability between the sexes fades. And the longer the races get, the more they overlap.

Sometimes, they even flip.

Poor Richard Ellsworth. In August 2019, he won the men's division in the Green Lakes Endurance Run in Fayetteville, New York, completing the fifty-kilometer course in just over four hours. But no trophy awaited him at the finish line.

Assuming that the overall first-place finisher would be a guy, officials planned to present one trophy to him and another to the fastest woman. Ellie Pell had other plans. She finished eight minutes before Ellsworth and took both trophies home. It was not the first time something like this had happened.

In 2002, Pam Reed was the overall winner of the grueling Badwater Ultramarathon, a 135-mile race through Death Valley cruelly run in the month of July. The next year, she did it again. In 2017, Courtney Dauwalter won the Moab 240 ultramarathon, running the course through the red rock canyons of Utah in two days, nine hours, and fifty-nine minutes. The second-place finisher—a man—crossed the finish line ten hours later. In January 2019, Jasmin Paris won the 268-mile Montane Spine Race in the U.K. in eighty-three hours, twelve minutes, and twenty-three seconds. Stopping at four rest stations along the way to pump breast milk for a fourteen-month-old daughter waiting for her at home, she still beat the course record by twelve hours. And Camille Herron has repeatedly won fifty-kilometer and hundred-kilometer ultramarathons outright.

The gap between elite men and women is closing, particularly in endurance athletics. Some studies show that women's leg muscles tend to be more resistant to fatigue than men's. In sports that test endurance rather than strength and speed, women may have the advantage.

NEVERTHELESS, THE WRONGHEADED notion that female walking is compromised remains pervasive. Author Rebecca Solnit calls this a "hangover from *Genesis*." She suggests in her 2000 book *Wanderlust: A History of Walking* that walking is "related to both thinking and freedom" and that historically men have thought women "deserve less of each."

Holly Dunsworth, the University of Rhode Island anthropologist who challenged the obstetrical dilemma on empirical grounds, agrees. "In a culture that has been heavily influenced by interpretations of the book of *Genesis*," she wrote, "the OD [obstetrical dilemma] offers a refreshing, scientific explanation for the consequences of the Fall." However, she added, the OD may be a flawed hypothesis, but "difficult labor, dangerous childbirth, and . . . helpless babies are not Eve's fault, but evolution's."

CHAPTER 12

Gait Differences and What They Mean

High'st queen of state, Great Juno, comes;
I know her by her gait.

—*William Shakespeare*, The Tempest, *1610–1611*

My wife and I work at the same college, and I occasionally spot her striding across the campus green. I know her just from her walk, even when I'm too far away to see her face. The way each of us walks is unique and recognizable, whether it's John Wayne's slightly off-balance swagger, Dorothy's skip toward Oz, Mae West's exaggerated hip swivel, or the lope of Shaggy from *Scooby-Doo*.

This observation is more than anecdotal.

In 1977, Wesleyan University psychologists James Cutting and Lynn Kozlowski conducted the first experiment to test whether people could identify one another by the way they walk. They recorded individuals walking and then converted their bodies into a series of small lights, similar to the motion-capture technology used in Hollywood today. That way, study participants could not pick up on cues such as hair color or body shape. The researchers found that even when people were turned into a string of lights, their friends were pretty good at spotting them.

Since then, repeated studies have confirmed that we are skilled at recognizing friends and family members solely by the way they walk. As it turns out, regions of our brains are fine-tuned to accomplish this.

In her 2017 study, for example, Carina Hahn, now a social scientist at the National Institute of Standards and Technology in Maryland, had nineteen participants lie in an MRI and watch videos of familiar people coming toward them. A region of the brain just behind the participants' ears (the bilateral posterior superior temporal sulcus) activated when they recognized people from their walks. When the walkers were close enough for their faces to be recognized, a different area of the participants' brains lit up.

But the way a person walks signals more than their identity. We are skilled at detecting moods, intentions, and even personality traits from the way someone walks. Slumped shoulders and a plodding gait are recognized as sadness. A bounce in one's step communicates happiness. A loud stomp can signify anger. Research shows that these inferences are not just a matter of intuition.

However, people aren't 100 percent accurate in interpreting these cues, and some of us are better at it than others. A 2012 study out of Durham University in the U.K. found that we perceive others as adventurous, warm, trustworthy, neurotic, extroverted, or approachable by the way they walk, but that the walkers often don't think of themselves that way at all. It appears that the inferences we draw this way are sometimes wrong.

But, as it turns out, some of those who are particularly good at it are psychopaths. In a 2013 study, Angela Book, a psychologist at Brock University in Ontario, Canada, showed videos of undergraduates walking to forty-seven maximum-security prisoners and asked them to rank how vulnerable the walkers were on a scale of one to ten. The prisoners—especially those characterized as psychopaths—revealed in follow-up questioning that they used gait cues to identify those who were frail or otherwise vulnerable to being preyed upon. Given the same task, undergraduate students were blind to these cues.

The implications were chilling. As Book pointed out, Ted Bundy, who confessed to raping and killing thirty women and girls in the 1970s, once boasted that he could "tell a victim by the way she walked down the street, the tilt of her head, the manner in which she carried herself."

It makes evolutionary sense that all animals—including humans—would be fine-tuned to identify different species and different individuals within those species, and even to recognize their moods by the way they move.

Given the evidence that different hominin species walked differently from one another in the past, it would have been beneficial, and perhaps even a matter of life and death, to know whether a group of hominins foraging in the distance belonged to your species or another. Subtle gait cues might have helped with these identifications. And if their gait did reveal them to be from your own species, could you tell if they were friends and family or strangers? Knowing the answer could have been the difference between avoiding conflict and inviting it.

Discerning moods from individual gaits would also have been advantageous. Was our hunting party successful, or were they hanging their heads and trudging slowly? Was someone limping? Was the large male's posture submissive? That could indicate a change in leadership, as happens today in chimpanzee troops.

Gait and posture cues might have been a critical means of communication in preverbal hominins, one likely as important, or perhaps more important, in our predecessors as it is for us today.

AS IT HAPPENS, our gaits aren't the only things about walking that betray our identities.

"Footprints are like fingerprints," Omar Costilla-Reyes explained.

He greeted me at the Brain and Cognitive Sciences Complex on the Massachusetts Institute of Technology (MIT) campus one crisp autumn morning. Costilla-Reyes wore an unzipped gray hoodie covering an I ♥ NASA T-shirt. Originally from Toluca, a small city outside Mexico City, he received his Ph.D. at England's University of Manchester, where he developed an algorithm that identifies individuals by the footprints they leave behind.

Costilla-Reyes identified twenty-four ways in which footprints differ from one person to another. His algorithm, he explained, accurately identifies individuals by their footprints 99.3 percent of the time, the top-performing result to date in footstep recognition.

I was impressed but skeptical. Couldn't his algorithm be fooled by someone faking his walk, such as Keyser Söze at the end of *The Usual Suspects*? Maybe so, Costilla-Reyes said, but as more data are used to train the machine learning algorithms, even those things shouldn't be able to fool it.

I started imagining the floors of airports equipped with pressure sensors. TSA agents would no longer need to scrutinize passports and boarding passes. Governments could know, from such pressure sensors or by cameras that capture our gaits, who is coming and going.

To learn more, I called Rama Chellappa, a professor of engineering at the University of Maryland, whose expertise is gait recognition and machine learning. In 2000, he received a U.S. Defense Department grant to investigate walking as an identification tool, though in the last two decades the research community has by and large moved to facial recognition, deemed a superior approach.

"We all walk differently, but this is still an academic exercise," he said, noting that the wrong camera angle or variations in the walking surface or carrying a load are enough to alter gait and impact accuracy. Furthermore, individual gait signatures would be difficult to extract from a crowd. I recalled the (possibly apocryphal) story of American spies being trained to put pebbles or pennies in their shoes to slightly alter their gaits and avoid being recognized.

Back at MIT, Costilla-Reyes told me that, because of advances in vision and machine learning, facial recognition is cheaper and more effective than gait recognition but that combining walking and facial expression would work really well.

Meanwhile, gait analysis also has possibilities for health professionals, Costilla-Reyes said. One of the first symptoms to appear in dementia and Alzheimer's patients is a change in gait. This could be detected early if doctors' offices and/or nursing homes were equipped with pressure-sensing equipment.

But that's not all. In 2012, Marios Savvides, a computer engineer at Carnegie Mellon University, developed an app that allows a smartphone to recognize the gait of its owner. Tiny gyroscopes and accelerometers inside smartphones can detect subtle differences in how someone walks. Since everyone has a unique gait, the phone will remain locked if it doesn't recognize the speed and movement of the user. A similar app is in use by some Pentagon officials, and a version is expected to be commercially available by 2021.

WALKING HAS ALWAYS been about more than moving from one place to another. It is, and always has been, a social phenomenon. Today, we celebrate the solitary, cerebral walks of Thoreau, Wordsworth, and Darwin, but rarely in our evolutionary history has walking alone been a good idea. Until very recently, a lone, contemplative walk would have surely ended in the piercing grip of a leopard's jaw.

Often, we walk collectively, like a school of fish, and it seems likely that our ancestors did, too. Anecdotally, it has long been known that people walking together subconsciously coordinate their gaits, but that wasn't empirically demonstrated until 2007.

Ari Zivotofsky of the Ocular Motor and Visual Perception Laboratory at Bar Ilan University in Israel and his collaborator Jeffrey Hausdorff of Tel Aviv University and Tel Aviv Sourasky Medical Center invited fourteen middle-school girls to walk down a school hallway. They found that when walking in pairs, the girls synchronized their gaits. Wearing blinders that blocked their views of one another did not seem to change the results. Not surprisingly, synchronization was most easily achieved when the girls held hands. I couldn't help but think back to the 3.66-million-year-old Laetoli footprints that show clear evidence of gait synchronization. Maybe the *Australopithecus* individuals who made them were holding hands.

A year after Zivotofsky's study, another found that people walking on adjacent treadmills at the gym synchronize their gaits.

In 2018, Claire Chambers, a postdoctoral student in the University of Pennsylvania's neuroscience department, analyzed human gaits from nearly 350 videos posted to YouTube. She found evidence that, from London to Seoul and New York to Istanbul, people—even complete strangers—synchronize their strides.

Sometimes, however, synchronized walking comes at a cost.

AT JUST EIGHTEEN years old, Stephen King wrote his first book, *The Long Walk*. In it, one hundred teenage boys and young men line up at the Maine-Canada border and walk south at four miles per hour. If they go under this speed threshold, they are issued a warning. Three warnings and they are executed by soldiers riding along-side them. Crowds line the streets, cheering them on. The long walk ends when there is only one competitor left standing.

What makes *The Long Walk* so gripping to someone like me who studies walking is the speed threshold: 4 miles per hour. A cross-cultural study done by psychologists Robert Levine and Ara Noren-zayan on over 2,000 people from thirty-one different countries found that humans walking alone on flat city streets average almost exactly 3 miles per hour. The Irish and Dutch tend to be slightly faster (3.6 miles per hour), and Brazilians and Romanians stroll at a more casual pace (2.5 miles per hour).

A 2011 study by researchers at the Human Motion Institute in Munich, Germany, collected walking speeds on 358 people of different ages and reported an average of 2.8 miles per hour, gradually slowing as we age. At this speed (3 miles per hour), humans are remarkably efficient locomotors and can walk and walk and walk without exhaustion. If King had made the cutoff 3 miles per hour, it just wouldn't have been as compelling a story.

But the cost to move our bodies increases with speed. Those boys in King's novel were exhausted, mentally, emotionally, and, yes, physically. Having to sustain four miles per hour to stay alive is what makes *The Long Walk* horrifying.

There are many reasons, some cultural and some anatomical, for why people naturally walk at different speeds, but one of them involves basic principles of energetics. Try walking at your normal pace. Now accelerate and walk faster. It takes energy to do this. But if you slow to a snail's pace, it also takes energy to resist your body's preferred speed. Everyone has an optimal walking speed, so what happens when people with different optimal speeds walk together?

Picture two walkers, a fast one and a slow one. Does the slow one speed up and absorb all of the energetic costs, or does the fast one slow down and take on that burden? What happens in a large group of people who all have different optimal walking speeds? When the Beatles walked across Abbey Road, did Ringo absorb the entire energetic burden or just some of it? The answer appears to be that people tend to meet in the middle, subconsciously settling into an optimal speed that minimizes energy costs for the entire group.

However, a twist occurs when two walkers are romantically involved. Seattle Pacific University professor Cara Wall-Scheffler's study of American college students found that the male in a heterosexual relationship absorbs the entire cost. That may be chivalrous but not entirely fair from a physiological point of view. Wall-Scheffler, the researcher who discovered that wide hips help women carry loads, found that those hips also give women a wider range of optimal walking speeds than men have. When women slow down or speed up, they don't expend as much energy as men do.

Nevertheless, walking has always been something we do together. For 97 percent of our species' history, and for 99 percent of time that bipedal hominins have walked the Earth, we have been nomadic hunters and gatherers. We roamed the landscape, walking from one food source to the next. We established temporary camps, and when the resources were nearly exhausted, we packed up our few belongings and moved on together.

Some human populations, including the Hadza of Tanzania and the Tsimané of Bolivia, still live this way, but today most people live in permanent settlements and eat farm products. We drive cars and fly in airplanes. And many of our cities, where more than half of all humans live, are designed in a way that makes walking from place to place difficult or even dangerous. Walking—the thing that made us human—isn't nearly as common as it used to be.

"Yes, there was a time when everybody walked: they did it because they had no choice," author Geoff Nicholson wrote. "The moment they had a choice, they chose not to do it."

As a result, our health has suffered.

CHAPTER 13

Myokines and the Cost of Immobility

I have two doctors, my left leg and my right.

—*George Macaulay Trevelyan*, Walking, *1913*

I've recently come across headlines promoting "10 Reasons to Go for a Walk Right Now," "9 Surprising Health Benefits of Walking," and "Benefits of Walking: 15 Reasons to Walk." In her 2014 book *Move Your DNA*, biomechanist Katy Bowman wrote, "Walking is a superfood." But as someone who studies human evolution, I see this somewhat differently. Walking is our default. Throughout our history, if we wanted to eat, we had to walk. What's new is *not* walking.

Consider the effect that immobility has had on our bones.

Our skeleton consists of two different kinds of bone. One, called cortical or compact, is the thick outer shell of our bones. The other, called trabecular, located in the joints where our bones meet, is a network of thinner, spongy bone arranged like a honeycomb. Compared with our ape cousins, humans have less of both kinds. We rely more on our trabecular skeleton to absorb, like a sponge, the high-impact forces of bipedalism.

Why, then, would we have so little of it?

Habiba Chirchir, a biological anthropologist at Marshall University in West Virginia, used CT scans to calculate trabecular bone density in the skeletons of humans, apes, and fossil hominins. She found that chimpanzees, fossil *Australopithecus*, Neandertals, and even Pleistocene *Homo sapiens* had the same density of trabecular bone in their joints: 30 to 40 percent. But humans today have less—20 to 25 percent. This drop in bone density appeared to occur suddenly in the last 10,000 years. Habiba suggests this happened because we don't move around as much as our ancestors did.

Tim Ryan, a Pennsylvania State University anthropologist, agrees. His study of four human populations—two nomadic groups and two farming communities—found that nomadic people have denser bones than farmers. While diet might have something to do with this, most scientists agree that bones are less dense in people who don't move around as much. In fact, humans have lost as much bone density in the last 10,000 years as an astronaut loses in a trip through the low-gravity conditions of space.

As we age, our bones naturally thin as levels of bone-stimulating estrogens fall, especially in postmenopausal women. But because we already have such low-density bone, this further reduction can lead to osteoporosis and broken bones in our aging population.

But osteoporosis might be the least of our worries.

WHEN I TURNED forty, my brother said, "Welcome to the back nine," a golf reference implying that I had reached the second half of my expected life span. It made me wonder what I could do to live longer and healthier.

The answer, says Steven Moore of the National Cancer Institute in Bethesda, Maryland, could be as simple as a daily walk. He and his research team compiled a decade's worth of data on 650,000 people and found that those who did the exercise equivalent of a twenty-five-minute daily walk—as long as they were not obese—lived close to four years longer than their more immobile counterparts. Even a ten-minute daily walk could make a two-year difference in life span.

Researchers at the University of Cambridge attempted to tease apart weight and inactivity as risk factors for an early death. They examined over 300,000 Europeans and discovered that inactivity caused twice as many deaths as obesity. They found that a twenty-minute daily walk cut the risk of dying by one-third. University of Copenhagen physiologist Bente Klarlund Pedersen explained in a 2012 TED Talk that it was "better to be fit and fat than lean and lazy."

To understand, we have to dive deeper into the science of physiology.

AFTER A BRIEF venture into astrophysics as a Cornell University undergrad, I switched majors and graduated with a degree in physiology. Instead of studying galaxies, I learned about bodies. The inner workings of living organisms bustle like rush hour at New York's Grand Central Terminal. The molecules in our bodies are in constant motion, arriving and departing in a steady flow. Sometimes they meet with handshakes or hugs. Sometimes they rush past one another without so much as a glance. Some carry gifts. Some carry guns. The complex dance of molecules is somehow both chaotic and orderly.

Walking affects this dance in significant ways. A good example and where a lot of the research on this topic has been done is breast cancer, especially a form called estrogen-receptor-positive breast cancer, which occurs in two-thirds of all cases. Breast cancer is exceedingly complex, but here are the basics.

Estrogen circulating in the bloodstream causes the cells of breast tissue to grow and divide as part of the normal physiology of a woman. Every time a cell divides, it copies its DNA, and every time it does so, there is a chance of a mistake—a mutation. Usually, this is not a big deal, but if a mutation occurs in a gene that limits how fast cells grow and divide, uncontrollable growth can produce a cluster of cells called a tumor. A mutation in a gene that keeps those cells in the breast where they belong could cause some to stray into the bloodstream and settle in the lungs, liver, bone, or brain. This process is called metastasis, and the result is stage IV breast cancer.

One in eight American women will be diagnosed with breast cancer in their lifetimes. Nearly 3,000 men are diagnosed annually as well. It kills 40,000 Americans and over half a million people globally every year.

But a daily walk reduces the chances of developing breast cancer. How? One possible explanation is that exercise lowers the levels of estrogen circulating in the blood. Anne McTiernan's team at the Fred Hutchinson Cancer Research Center in Seattle showed in 2016 that exercise increases the body's production of a molecule scientists call sex hormone binding globulin. This molecule attaches itself to estrogen, reducing its concentration in the blood by 10 to 15 percent, thereby reducing the chance of a mutation in the DNA of breast tissue.

Even if a mutation occurs, exercise appears to help the damaged DNA repair itself. Study participants who exercised at least twenty minutes a day had a slightly (1.6 percent) better ability to repair DNA copying mistakes, although it's unclear how this works.

If the copying mistakes don't get fixed and cancer results, walking still helps. In a study of nearly 5,000 women diagnosed with breast cancer, Crystal Holick and her former colleagues from the Fred Hutchinson center found that exercise—even just an hour of walking a week—decreased the chances of dying by about 40 percent. A follow-up by Saudi Arabian cancer researchers Ezzeldin Ibrahim and Abdelaziz Al-Homaidh put the number at 50 percent for estrogen-positive breast cancer. They also found that exercise reduced the chance of cancer recurring after remission by 24 percent. Similar reductions in recurrence have been found in men with prostate cancer who routinely walk after diagnosis. In fact, a 2016 study of nearly 1.5 million people found that moderate exercise lowers the risk of developing thirteen different cancers.

While cancer claims too many lives, the number one killer in industrialized nations is cardiovascular disease. In its various forms, it is responsible for one in four deaths, or 600,000 Americans a year. Walking can help stave that off, too. Frequent walkers have lower heart rates and lower blood pressure than sedentary individuals. A 2002 study of just under 40,000 American men found that a daily thirty-minute walk lowered the risk for coronary heart disease by 18 percent.

Coronary heart disease is all but unheard of among hunter-gatherers. Dave Raichlen, a professor of human biology at the University of Southern California, reported that the Hadza of northern Tanzania are fourteen times more active than the average American. They also have lower blood pressure as they age, lower cholesterol, and not a hint of cardiovascular disease. A study of Bolivia's Tsimané people similarly found them to have low levels of coronary heart disease and five times less blockage of their arteries than the average person in the industrialized world.

Diet has a lot to do with this, but there is evidence that physical activity plays a critical role. Perhaps, though, not in the way you may think.

DUKE UNIVERSITY ANTHROPOLOGIST Herman Pontzer has spent the last decade trying to understand how the human body uses energy. He traveled to northern Tanzania and lived with the Hadza, collecting data on how much they move and how much energy they expend. Everyone, including Pontzer, figured that the Hadza used more energy than the typical American. After all, Hadza adults walk between six and nine miles a day while the average American, according to Nielsen Media Research, spends six hours a day staring at screens.

But what Pontzer found was shocking and forces us to think about our bodies in a different way. The total daily energy used by active Hadza people and couch-potato Americans is the same.

How is this possible?

A clue is hidden in the one thing that walking does *not* help us do: lose weight. It turns out that humans are so efficient at walking that a 150-pound person would have to walk at least *seventy* miles to lose a pound. So the extra steps the Hadza take compared to typical Americans don't burn much more energy. But the Hadza don't just walk. They dig, climb, and run. Surely, they should be using more energy.

The currently accepted hypothesis for this mystery, Pontzer told me, is that human bodies all over the world have the same daily energy allowance. How they spend that energy varies from culture to culture and from person to person. The Hadza use energy getting from one place to another, gathering food, fighting off illnesses, carrying children and growing new ones. Americans do many of those same things, but because we aren't as active, our bodies spend the excess energy on something else: ramping up our body's inflammation response.

Here's why that is a health problem.

The inflammation response refers to the way our bodies recruit large, vigilant, amoeba-looking cells called macrophages to ward off infections or repair injuries. These cells, whose name means "big eaters," are key components of our immune system. They make an infection-fighting protein called tumor necrosis factor (TNF). Among a variety of roles in the body, TNF tells the hypothalamus to crank up the body temperature when we are overrun by a virus or bacteria—something, of course, called a fever.

But chronically high levels of TNF have been linked to heart disease.

In 2017, Stoyan Dimitrov of the University of Tübingen in Germany discovered that walking can downshift the production of TNF. In fact, it dropped by 5 percent after a brisk twenty-minute walk.

How?

The answer appears to involve a whole class of proteins that weren't even in the textbooks when I studied physiology as an undergraduate.

IN THE LATE 1990s, a research team led by Danish physiologist Bente Klarlund Pedersen got interested in a protein called interleukin-6, which white blood cells use to communicate with one another. They discovered that interleukin-6 levels in marathon runners are a hundred times higher at the end of a race than at the start.

To figure out what was going on, Pedersen strapped weights to the ankles of six men who remained seated during the experiment. An IV was placed in each leg so that blood could be drawn. Every few seconds, the men extended one leg forward in a slow kick while keeping the other leg still. The concentration of interleukin-6 went up in blood drawn from the exercising leg but not in the other one. Pedersen surmised that the muscles themselves were making interleukin-6 and releasing it into the bloodstream.

This was a revolutionary idea.

Many organs in our bodies make molecules and release them into our bloodstream as a way to talk with other organs. These endocrine organs include the pancreas, the pituitary gland, the ovaries, and the testes. But few had thought of *muscle* as an endocrine organ until Pedersen's work. Interleukin-6 was just the start. Scientists have now discovered over a hundred molecules that our muscles make and release into the blood as we walk. Pedersen's team discovered that one of these, oncostatin M, shrank breast tissue tumors in mice and could be yet another reason why exercise is beneficial to humans with breast cancer.

In 2003, Pedersen coined a name for this amazing family of molecules: myokines.

As a myokine, interleukin-6 is an anti-inflammatory. Among other roles, it helps shut down the problematic tumor necrosis factor (TNF). It is the body's natural ibuprofen. Pedersen's team also discovered that interleukin-6 can mobilize cells called "natural killers" to attack and destroy cancerous tumors, at least in mice.

For some reason, this myokine needs to be produced by muscles during exercise in order to work. But that does not require walking. Can the 3 million Americans in wheelchairs generate myokines? Yes. Researchers at the Department of Rehabilitation Medicine at Wakayama Medical University in Japan have discovered elevated interleukin-6 levels, and lowered tumor necrosis factor, after wheelchair half-marathons and basketball games. As Juliette Rizzo, 2005 Ms. Wheelchair America, said, "Walking is a way to get from A to B, and I do that."

Myokines, however, are not magic potions. They cannot be injected or swallowed. They are made only when the body is in motion, and in modern societies it often is not. On average, Americans take 5,117 steps a day, which is a third of what the average Hadza takes. Do we have to walk that much to be healthy? How much should we walk to ward off heart disease, certain cancers, and type 2 diabetes?

According to my smartphone, the answer is 10,000 steps a day. If I take that many, the pedometer app on my phone registers its approval by changing color from a disappointed red or orange to a happy green. Where does this magical 10,000-step threshold come from? To find out, we must travel back in time to the 1964 Summer Olympic Games in Japan.

In Tokyo that year, Abebe Bikila of Ethiopia defended his gold medal in the marathon, setting a new world record in 2:12:11.2. American sprinter and future NFL hall of famer Bob Hayes raced a hundred meters down a cinder track in only 10.06 seconds, tying the world record. Joe Frazier slugged his way to gold in boxing. And Soviet gymnast Larisa Latynina, competing in her last Olympics, took home six more medals, bringing her total to eighteen and making her the most decorated of any Olympian until American Michael Phelps swam along.

The Olympics inspired the people of Japan. For the first time, the games were broadcast live on television, and by 1964, 90 percent of Japanese homes had a TV. In this, Yoshiro Hatano, a professor of health and welfare at Kyushu University, saw opportunity. He was concerned with how sedentary the Japanese public had become and with the rising prevalence of obesity in his home country. His research on walking showed that people there took between 3,500 and 5,000 steps per day. By his calculation, that was not enough to be healthy.

The following year, Hatano worked with watchmaker Yamasa Tokei to create a device that hooked onto people's waists and counted the steps they took. They called it Manpo-kei. In Japanese, *man* means "10,000," *po* is "step," and *kei* is "meter." Thus, the 10,000-step meter.

My smartphone walking app defaults to 10,000 steps as a daily goal. Most Fitbits do the same. Although the 10,000-step goal was based on some research by Hatano, it was mostly a marketing gimmick. Yet, over half a century later, it is still with us. But is this a meaningful number? How many daily steps should we take?

From 2011 to 2015, I-Min Lee, an epidemiologist at Brigham and Women's Hospital in Boston, asked nearly 17,000 women whose age averaged seventy-two years to wear an accelerometer for a week. As a group, they averaged 5,499 steps a day, a shade more than the number a typical American adult takes.

In the slightly over four years that followed, 504 of these women died. Lee found that the number of daily steps the study participants took was a good predictor for who was still alive and who was not. She found that women who averaged at least 4,400 steps fared much better than those who took only 2,700. With daily steps up to 7,500, women continued to be better off than those taking fewer steps. But that's where it plateaued. Going beyond 7,500 steps did not make a difference.

But for a younger population, the plateau may not occur at 7,500 steps. The amount of walking required to yield health benefits depends on age and activity level. Keeping it simple, Lee recommends that everyone try to take 2,000 more steps a day than they currently average.

One way to add that many steps to your routine is to get a dog.

Dogs were the first animals domesticated by our species. Ancient DNA from a canine's rib found in Siberia reveals that humans and the wolf ancestors of dogs started hanging out together by 30,000 years ago. In comparison, pigs and cows were domesticated closer to

10,000 years ago. As humans migrated around the globe, our dogs walked alongside us.

Even today, dog owners average almost 3,000 more daily steps than non–dog owners and are more likely to reach the recommended 150 minutes of walking per week.

IN ADDITION TO preventing some cancers and reducing the risk of dying from cardiovascular disease, a daily walk can prevent auto-immune diseases and can help ward off type 2 diabetes by lowering blood sugar levels. It improves sleep and lowers blood pressure. It decreases circulating cortisol levels, which helps reduce stress. In a study of almost 40,000 women over forty-five years old, a thirty-minute daily walk reduced the risk of stroke by 27 percent. Despite these health benefits and admirable attempts to get our sedentary population on its feet, it is an uphill battle. Many futurists predict that our walking days are behind us.

In Kurt Vonnegut's novel *Galápagos*, our descendants a million years in the future evolve aquatic adaptations. They lose the ability to walk and become streamlined for swimming. The Pixar animated movie *WALL-E* doesn't go that far, but it, too, predicts that future humans, confined to lounge chairs aboard the *Axiom* spaceship with robots attending to their every need, won't walk.

Could it be that we humans will stop doing the very thing that defined us from the beginning?

I sure hope not for the sake of our physical and, it turns out, our mental health.

CHAPTER 14

Why Walking
Helps Us Think

Moreover, you must walk like a camel, which is said to be the
only beast which ruminates when walking.

—Henry David Thoreau, "Walking," 1861

C harles Darwin was an introvert. Granted, he spent almost five
years traveling the world on the *Beagle* recording observations
that produced some of the most important scientific insights ever
made. But he was in his twenties then, embarking on a privileged,
nineteenth-century naturalist's version of backpacking around Europe during a gap year. After returning home in 1836, he never again
stepped foot outside the British Isles.

He avoided conferences, parties, and large gatherings. They made
him anxious and exacerbated an illness that plagued much of his
adult life. Instead, he passed his days at Down House, his quiet home
almost twenty miles southeast of London, doing most of his writing
in the study. He occasionally entertained a visitor or two but preferred to correspond with the world by letter. He installed a mirror
in his study so he could glance up from his work to see the mailman
coming up the road—the nineteenth-century version of hitting the
REFRESH button on email.

Darwin's best thinking, however, was not done in his study. It was done outside, on a lowercase *d*–shaped path on the edge of his property. Darwin called it the Sandwalk. Today, it is known as Darwin's thinking path. Janet Browne, author of a two-volume biography of Darwin, wrote:

> *As a businesslike man, he would pile up a mound of flints at the turn of the path and knock one away every time he passed to ensure he made a predetermined number of circuits without having to interrupt his train of thought. Five turns around the path amounted to half a mile or so. The Sandwalk was where he pondered. In this soothing routine, a sense of place became preeminent in Darwin's science. It shaped his identity as a thinker.*

Darwin circled the Sandwalk as he developed his theory of evolution by means of natural selection. He walked to ponder the mechanism of movement in climbing plants and to imagine what wonders pollinated the fantastically shaped and colorful orchids he described. He walked as he developed his theory of sexual selection and as he accumulated the evidence for human ancestry. His final walks were done with his wife Emma as he thought about earthworms and their role in gradually remodeling the soil.

In February 2019, I had the meta-experience of walking Darwin's thinking path to think about how walking helps you think. It was school vacation in London, and I had to compete with families arriving in droves to see where Darwin had lived and worked. The desk in his study is still cluttered with books, letters, and small specimen boxes containing pinned insects. Hanging from a nearby chair is his black jacket, black bowler hat, and a wooden walking stick. The stick has a helical design like a crawling tendril and looks freshly polished. The bottom of the walking stick, however, is well worn—evidence of miles on the Sandwalk.

I walked out the back kitchen of the cream-colored home, passed the green trellis and vine-covered columns holding up Darwin's back porch, crossed the beautifully groomed garden, and entered the Sandwalk. I was alone. The day was cool and blustery. Gray clouds hung low on the horizon and moved swiftly overhead, dropping an intermittent drizzle. Occasional breaks in the clouds allowed the sun to peek through, making the raindrops flicker.

I could hear planes from the nearby London Biggin Hill Airport and the hum of a lorry traveling along A233. But those modern sounds were fleeting. It was easy to imagine that it was 1871 and that I was taking a walk with Darwin himself. I could hear the chatter of gray squirrels but tuned them out as well since they are an invasive North American species introduced into England in 1876.

I stacked five flat flints at the entrance for the five laps I would take and began my walk, first along the meadow and then counterclockwise into the woods. The Sandwalk is alive. Starlings and crows fly overhead, filling the air with their trills and gurgles. Ivy inches up the thick trunks of alder and oak trees toward the sunlight. Underfoot, fungi decompose wet leaves, emitting the smell of fresh earth. I picked up a clump of cockleburs just off the path, and their hooks pulled on the folds of my hand and latched to my jacket. With each step the gravel crunched, and my shoes occasionally slipped on damp stones made smooth by thousands of footsteps, including some taken by Darwin himself.

Down House is not a place of magic, nor is it a place of worship. Looping the Sandwalk one flint at a time did not endow me with the wisdom to continue my scientific pursuits. It turns out, any walk outdoors has the potential to unlock our brains. The Sandwalk just happened to be where the unlocking of one nineteenth-century brain helped change the world and our place in it.

But why? Why does walking help us think?

YOU ARE UNDOUBTEDLY familiar with this situation: You're struggling with a problem—a tough work or school assignment, a complicated relationship, the prospects of a career change—and you cannot figure out what to do. So you decide to take a walk, and somewhere along that trek, the answer comes to you.

The nineteenth-century English poet William Wordsworth is said to have walked 180,000 miles in his life. Surely on one of those walks he discovered his dancing daffodils. French philosopher Jean-Jacques Rousseau once said, "There is something about walking which stimulates and enlivens my thoughts. When I stay in one place I can hardly think at all; my body has to be on the move to set my mind going." Ralph Waldo Emerson's and Henry David Thoreau's walks in the New England woods inspired their writing, including "Walking," Thoreau's treatise on the subject. John Muir, Jonathan Swift, Immanuel Kant, Beethoven, and Friedrich Nietzsche were obsessive walkers. Nietzsche, who walked with his notebook every day between 11 a.m. and 1 p.m., said, "All truly great thoughts are conceived by walking." Charles Dickens preferred to take long walks though London at night. "The road was so lonely in the night, that I fell asleep to the monotonous sound of my own feet, doing their regular four miles an hour," Dickens wrote. "Mile after mile I walked, without the slightest sense of exertion, dozing heavily and dreaming constantly." More recently, walks became an important part of the creative process of Apple cofounder Steve Jobs.

It is important to pause and reflect on these famous walkers.

They are all guys. Little has been written about famous women who regularly walked. Virginia Woolf is one exception. Simone de Beauvoir is another. More recently, Robyn Davidson trekked with her dog and four camels across Australia and wrote about it in her book *Tracks*. In 1999, Dorris Haddock, an eighty-nine-year-old grandmother from Dublin, New Hampshire, walked 3,200 miles from coast to coast to protest United States campaign finance laws.

Historically, however, walking has been the privilege of white men. Black men were likely to be arrested, or worse. Women just out for a walk were harassed, or worse. And, of course, rarely in our evolutionary history was it safe for anyone to walk alone.

Perhaps it is a coincidence that so many great thinkers were obsessive walkers. There could be just as many brilliant thinkers who never walked. Did William Shakespeare, Jane Austen, or Toni Morrison walk every day? What about Frederick Douglass, Marie Curie, or Isaac Newton? Surely the astoundingly brilliant Stephen Hawking did not walk after ALS paralyzed him. So walking is not essential to thinking, but it certainly helps.

MARILY OPPEZZO, A Stanford University psychologist, used to walk around campus with her Ph.D. advisor to discuss lab results and brainstorm new projects. One day they came up with an experiment to look at the effects of walking on creative thinking. Was there something to the age-old idea that walking and thinking are linked?

Oppezzo designed an elegant experiment. A group of Stanford students were asked to list as many creative uses for common objects as they could. A Frisbee, for example, can be used as a dog toy, but it can also be used as a hat, a plate, a bird bath, or a small shovel. The more novel uses a student listed, the higher the creativity score.

Half the students sat for an hour before they were given their test. The others walked on a treadmill.

The results were staggering. Creativity scores improved by 60 percent after a walk.

A few years earlier, Michelle Voss, a University of Iowa psychology professor, studied the effects of walking on brain connectivity. She recruited sixty-five couch-potato volunteers aged fifty-five to eighty and imaged their brains in an MRI machine. For the next year, half of her volunteers took forty-minute walks three times a week. The other participants kept spending their days watching *Golden Girls* reruns (no judgment here; I love Dorothy and Blanche) and only participated in stretching exercises as a control. After a year, Voss put everyone back in the MRI machine and imaged their brains again. Not much had happened to the control group, but the walkers had significantly improved connectivity in regions of the brain understood to play an important role in our ability to think creatively.

Walking changes our brains, and it impacts not only creativity, but also memory.

In 2004, Jennifer Weuve of Boston University's School of Public Health studied the relationship between walking and cognitive decline in 18,766 women aged seventy to eighty-one. Her team asked them to name as many animals as they could in one minute. Those who walked regularly recalled more penguins, pandas, and pangolins than the women who were less mobile. Weuve then read a series of numbers and asked the women to repeat them in reverse order. Those who walked regularly performed the task much better than those who didn't. Even walking as little as ninety minutes per week, Weuve found, reduced the rate at which cognition declined over time. Therefore, because cognitive decline is what occurs in the earliest stages of dementia, walking might ward off that neurode-generative condition.

But correlation does not equal causation. Otherwise, one could interpret graveyards as places where giant stones fall from the sky and kill unsuspecting, mostly elderly, people. Perhaps the arrow of causality was pointing in the wrong direction. Maybe mentally active people were simply more likely to go for a walk. Researchers had to dive deeper.

For that, let's visit the gross anatomy lab where my students dissect human cadavers.

BY AUGUST, MY students have already spent eight intense weeks exploring every inch of the insides of individuals who have donated their bodies for dissection by Dartmouth College premeds. They have teased tissues apart to find sinewy heart valves and calcified arteries. They have followed blood vessels that make loops as ill conceived as the roads around Boston. The serious stillness of the dissection room changes to an exhilarating buzz when the students discover a hip replacement or the mesh of a stent. The reaction is less celebratory when a student discovers a cancerous growth or accidentally nicks the lower bowel.

After each dissection, the students return the organs to their proper places and fold the layers of tissue and skin back as if they are closing the pages of a sacred text. I've seen students gently hold their first patient's hands while their classmates slice into paper-thin skin. Cadavers are the best teachers my students will ever have.

Surprised that someone who studies fossils also teaches anatomy to premed students? Don't be. Paleontologists know anatomy. A fossil could be from any of the more than two hundred different bones in the body and from one of dozens of different kinds of animals that lived on the landscape. When I pick up a fossil, I need to quickly determine what bone it could be. Is it a humerus? A vertebra? Part of a jaw? Is it from an ancient antelope or a zebra? A monkey or an early human? Little bumps on a fossil bone hold clues to which muscles and ligaments anchored there in life. Some bones have grooves and holes that would have allowed for the passage of blood vessels and nerves millions of years ago when the hearts of those ancient animals were still beating. All of this requires a knowledge of anatomy, and that means many hours in the dissection lab.

In week nine, the saws come out. Weeks of delicately nudging tissues aside to identify muscles, nerves, and blood vessels give way to the brutal act of extracting the brain. Sawing off the top of a skull upsets many students, and it should. It is a very unnatural act. When the buzzing electric saws are turned off, the room falls silent. Few students talk, and none make jokes. A singeing smell, reminiscent of burnt hair, lingers in the air. Sometimes a hammer and chisel are required to crack areas too thick for the saw to penetrate.

Students generally express emotion when they hold a heart. They are awed when they hold the brain. The brain *is* the person. My students are often surprised by how light and spongy it is—and how vulnerable. They run their fingers over the folds and into the grooves. With a large knife, one of my students cuts through the brain as if it were a cantaloupe, splitting it into equal left and right halves. There, sitting on top of the brain stem, is a thick loop of tissue about the length of my pinky finger. To me, it looks like a gummy worm. To early anatomists, it looked like the tail of a seahorse, so they named it after the mythical Greek sea monster that had the body of a horse and the tail of a fish—the hippocampus. The hippocampus is the memory center of the brain. When the neurons in that small bit of brain were firing, they stored our cadaver's many memories.

Maybe in his last years he couldn't remember the name of his third-grade teacher anymore but could distinctly recall the shape and color of her glasses. Maybe he could still conjure the earthy smell of his childhood dog Sadie after a hike in the woods. He could ask his hippocampus to recall the exact moment the girl he secretly had a crush on for three years smiled at him as their high school English teacher, Mr. Austin, was reading the poem "Thanatopsis." He could still feel the velvety petals of the orchid she had in her hair on their wedding day. He could remember how many home runs Carl Yastrzemski hit in 1964, but on some days, he could not remember his wife's name. It puzzled and frustrated him. He would get angry. When he calmed, she held his hand, and he sang the entirety of "Smoke Gets in Your Eyes," their wedding song, to a woman whose name he had forgotten. Perhaps, on the day he died, he asked his son to turn on the Sox game and get Sadie from the backyard.

The pain and frustration of forgetting should make us want to do everything we can to maintain this region of our brain, the central storage facility for our memories. To be sure, other kinds of memories are stored elsewhere in the brain: the ability to recognize faces, so-called implicit memories such as how to ride a bike, and so-called explicit memories such as the date World War II began. But the hippocampus is the depository of our life stories.

As we get old, however, our brain gets smaller. In our later years, the hippocampus shrinks at a clip of 1 to 2 percent each year, and it becomes more and more difficult to recall things that used to come to us instantly. A colleague nearing retirement used to joke that it was taking longer for the little guy in his brain to shuffle through the file cabinet of memories to find what he was looking for. There

were just more files to go through, they weren't well organized anymore, and that little guy had to walk with a cane.

What are we to do about this?

Walk.

IN 2011, UNIVERSITY of Pittsburgh psychologists gathered 120
aging but otherwise healthy folks from the community. They gave
them MRIs and measured the size of their hippocampus. Then, half
of them were asked to walk forty minutes three times a week. The
other half just did stretches but did not take the long walks. After a
year, the stretching group had lost between 1 and 2 percent of the
volume of the hippocampus. That was expected. But something extraordinary happened with the walkers. Not only had they not lost
any hippocampal volume. They *gained* some. The walking group,
on average, had grown the hippocampus by 2 percent. Accordingly,
their memory had improved.

The hippocampus, it turns out, can regenerate, and even just a
daily walk can promote growth. Walking can not only delay some
effects of aging, but can reverse them. But how?

One explanation is that walking, or any exercise, helps get the
blood flowing, and, indeed, this happens. In 2018, Sophie Carter of
Liverpool John Moores University took MRI brain scans of people
who walked for two minutes every half hour or so and of others who
sat all day. She found that those who got up and walked around had
significantly greater blood flow in the middle cerebral and carotid
arteries. But blood is just the vehicle. It must be carrying something
of critical importance to the brain.

Myokines. Those molecules released by contracting muscles target the brain, and blood flow delivers them. One of those myokines
is called irisin, named after Iris, the Greek goddess of rainbows and

Hera's personal messenger. In 2019, researchers at the Federal University of Rio de Janeiro in Brazil found alarmingly low levels of irisin in humans with Alzheimer's, a disease that impacts one in ten people over sixty-five.

When the Brazilian researchers blocked the production of irisin in mice, our rodent cousins performed terribly trying to remember where the cheese was in a maze. When the irisin flowed again, the same mice recovered. The mice that performed the best were the ones that exercised. In mice, at least, irisin goes straight to the hippocampus, where it protects neurons from degeneration.

Another of these myokines is called brain-derived neurotrophic factor, or BDNF. It is not as fun to say as irisin, but it might be even more important. The walkers in the University of Pittsburgh study whose hippocampus increased by 2 percent also had higher levels of BDNF than the nonwalking group. John Ratey, a clinical psychiatry professor at Harvard Medical School, calls BDNF "Miracle-Gro for your brain."

But walking does not help just with the hippocampus and memory. There is some evidence that it helps relieve symptoms of depression and anxiety.

"I told myself I wasn't doing any walking because I was so depressed and enervated," British author Geoff Nicholson wrote in *The Lost Art of Walking*. "And then I thought of something. Perhaps I was depressed and enervated precisely because I wasn't doing any walking."

Those who struggle with depression describe it as an exhausting abyss of hopelessness. When you are in it, you feel like you'll never get out. One in twelve Americans know this feeling. While many studies have shown that a regular walk can alleviate symptoms of depression and anxiety, it doesn't work for everybody. Furthermore, the benefits appear to depend on where you walk. To understand why, we need to return to the anatomy lab and look again at the brain.

To the untrained eye, the folds and fissures of the brain seem randomly distributed. To a neurologist, they are a map that reveals the workings of our most magnificent organ. The folds in the back of the brain are where visual cues are processed. A strip of nervous tissue across the top helps coordinate movement. The bulge in the front of the brain is where we make plans. Fifty-two different regions of the brain were identified and named in the early twentieth century by German neurologist Korbinian Brodmann, and each one now bears his name. Brodmann area 22, for example, processes sounds. Brodmann 44 and 45 help you talk.

About three inches behind the bridge of your nose is Brodmann 25, or what modern neurologists call the subgenual prefrontal cortex (sgPFC). It plays an important role in regulating our moods, displaying increased activity during periods of sadness and rumination.

Around the time Marily Oppezzo was asking her Stanford students how many uses they could list for a Frisbee, fellow Cardinal Greg Bratman was wondering how a walk in the woods could improve your mood. Bratman, then a doctoral student interested in the intersection of environment and psychology, had thirty-eight people fill out a survey that included questions about their mood and negative self-reflection. He was particularly interested in whether a problem was eating away at some of them. The survey was tallied into what was called a rumination score. Bratman then took MRI scans of the sgPFC, assessing the blood flow to this region. Then he sent the participants out for a walk.

Half took a three-and-a-half-mile walk through greenspace on the Stanford campus. There was fresh air, the shade of coast live oak trees, and screeches from western scrub jays. The other half walked the same distance on a sidewalk along El Camino Real, a busy, multilane street through the heart of Palo Alto. There, they had to be alert for cars coming and going from gas stations, hotels, parking lots, and fast-food restaurants. When the participants returned, they were given another survey and another MRI scan.

Those who had walked along the busy road showed no change in their rumination scores or in the blood flow to the sgPFC. But those who had walked in the woods had lowered their rumination scores, and blood flow to the sgPFC had reduced significantly.

For our mental health, it seems, we should walk where there are trees, birds, and the soft whispers of the wind.

That brings us back to the other Stanford study, the one by Marily Oppezzo. After her treadmill walkers did better than nonwalkers on their creativity tests, she added a test group that took its walk outside. The ones on footpaths performed even better than the treadmill walkers at coming up with original ideas.

Unfortunately, we are not only walking less but, with so many of us living in urban areas, we are doing it in places that erase some of the health benefits.

PERHAPS RAY BRADBURY was right about what the future may hold.

In his 1951 short story "The Pedestrian," set a hundred years in the future, a writer named Leonard Mead goes for his nightly walk. Bradbury writes:

> To enter out into that silence that was the city at eight o'clock of a misty evening in November, to put your feet upon that buckling

concrete walk, to step over grassy seams and make your way, hands
in pockets, through the silences, that was what Mr. Leonard Mead
most dearly loved to do.

As usual, Mead walks alone as his city neighbors watch TV, their
windows all aglow. A robotic police officer stops him and asks what
he is doing.

"Just walking," he responds.

"Walking where? For what?" the police officer wants to know.

"For air, and to see, and just to walk," he responds.

"Have you done this often?"

"Every night for years," Mead says.

"Get in," orders the officer.

The story ends with Mead in the back of a police car that is taking him to the Psychiatric Center for Research on Regressive Tendencies.

CHAPTER 15

Of Ostrich Feet and Knee Replacements

Time wounds all heels.

—*Groucho Marx*, Go West, *1940*

I choose to walk at all risks.

—*Elizabeth Barrett Browning*, Aurora Leigh, *1856*

In 1490, Leonardo da Vinci drew the Vitruvian Man, a sketch of a man with arms and legs stretched to the rims of a bounding circle and square. It was created in part to demonstrate the ideal proportions of the human form as surmised by the first-century Roman architect Vitruvius, but the Vitruvian Man is far from ideal. In fact, this icon displays one of the scars of our evolutionary past.

Dr. Hutan Ashrafian, a lecturer at Imperial College in London, noticed in 2011 that Vitruvian Man possesses an odd bulge just above the left side of his groin. He recognized it at once as an inguinal hernia, something more than a quarter of all men develop in their lifetimes. If left untreated, an inguinal hernia can be deadly—as the cadaver that formed the basis for Leonardo's drawing demonstrates.

Inguinal hernias are a direct result of bipedalism.

At birth, human testes are positioned within the male abdomen near the organs of the urinary system, but in the first year of life, they migrate through the abdominal cavity and into the scrotum. This migration creates something called the inguinal canal, a weak spot in the abdominal wall. This happens in many other mammals, too, but without negative consequences. Because we are upright, however, gravity tugs our insides down, and sometimes our intestines squeeze through the inguinal canal and become choked off, resulting in a dangerous and occasionally fatal condition.

This odd route that testes take in most mammals is a byproduct of developmental constraints and a deep, evolutionary history. The testes remain internal in some mammals, including dolphins, elephants, and armadillos, but because lower temperatures are required for normal sperm functioning in most mammals, their testes hang away from the body where it's cooler. Fish, however, also have internal testes. Because mammals share a common ancestor with them—a fish that lived upward of 375 million years ago—our testes retain a developmental starting point in the abdomen as a vestige of our aquatic past.

Although walking has physical and mental benefits, it also has downsides. In part, this is because we were not created from scratch. We are modified apes. Our lineage has had 6 million years or so to fine-tune our bodies to upright walking, but evolution does not create perfection. Instead, it results in forms that are just good enough to survive, reproduce, and carry on the lineage. The fossil record is littered with extinct animals that once were well adapted for life but died off when the inevitable environmental change pulled the rug out from under them. Even the best-adapted survivors, including humans, are junkyards of preexisting forms, modified by the hands of natural selection and rich with the echoes of the past.

If we had been designed as bipeds from scratch, perhaps we might look something like CASSIE.

"In the future, robots will be able to do everything that humans do, but better," Jonathan Hurst, an Oregon State University professor of mechanical engineering and robotics, told me when I visited his lab. He imagines a not-so-distant future in which bipedal robots deliver packages, serve meals, and help with search-and-rescue missions.

Unconvinced that bipedalism is the best way to move around the world, I asked why Hurst designs two-legged robots. Why not make them quadrupedal? Hell, why not just give them wheels? Robots, Hurst told me, will be moving in a world designed for humans, so it makes sense for them to move the way we do.

But Hurst's designs do not look human.

I met CASSIE in February 2019. Hurst's students led the robot onto a treadmill, and it marched at a steady three miles per hour, the average speed of a person, on small, padded feet. CASSIE doesn't look anything like C-3PO, Bender, the Terminator, or Johnny 5. It isn't humanoid at all. Instead, the four-foot-tall, seventy-pound CASSIE is all legs. But unlike my straight legs, CASSIE's are thin and bent, and the motors powering the robot are near the hips. I've seen this design before—in large terrestrial birds. In fact, CASSIE is short for cassowary, a hundred-pound flightless bird from New Guinea.

But Hurst did not intentionally model CASSIE on any living animal. For the last two decades, his research team has studied the physics behind bipedalism—what Hurst calls the "universal truths" of walking. These principles, rather than any preconceived design, have guided the development of his robots, and they don't look like us.

ANOTHER PROBLEM HANDED down to us by the vestiges of our evolutionary past is one that impacts women far more than men. To understand it, we have to travel back 30 million years.

Just as humans did not evolve from chimpanzees, apes did not evolve from monkeys. Instead, they share a common ancestor. Paleontologists working in 30-million-year-old North African sediments have discovered what this common ancestor looked like. It was a cat-size primate called *Aegyptopithecus*. The name means the ape from Egypt, but, of course, *Aegyptopithecus* was not an ape. It had the teeth of an ape, but it moved on all fours like a monkey. And, unlike an ape, *Aegyptopithecus* had a long tail.

Over the next 10 million years, this lineage began to split into two forms. One retained the tail, evolved differently shaped teeth, and diversified into today's African and Asian monkeys. The other lost the tail and eventually diversified into today's apes. Apes, a group of primates that includes gibbons, orangutans, gorillas, chimpanzees, bonobos, and humans, do not have tails but retain the muscles that once moved a tail.

These muscles still anchor to our vestigial tailbone, or coccyx, and they create a sling of muscle that forms the floor of our pelvis. The same muscles that wag a dog's tail and allow some monkeys to hang by their tails have been repurposed in the apes to support the internal organs against the pull of gravity. But, in upright walking humans, the pull of gravity is sometimes too much for this sheet of muscle.

Hence, pelvic organ prolapse, the debilitating condition in which internal organs sometimes protrude into the vagina.

THE MUSEUM OF Science in Boston houses a large number of live animals—some injured wildlife, some confiscated pets—that are used to teach lessons about ecology, animal behavior, and evolution. One of my favorites when I taught there was Alex, a small American alligator. Alex was only a few years old and weighed no more than ten pounds, but he drew quite an audience whenever I took him from his tank so visitors could see him up close.

Alex was good-tempered, but I always knew when he was agitated. The muscles at the base of his tail would tense just before he was about to flail to escape my grip. When I felt his tail muscles clench, I simply tipped him upright, his head facing the ceiling and his tail the floor. That drained some blood from his head, and it calmed him. After a few seconds, I could continue teaching about reptiles.

Alligators have valves in their veins, but they appear to be only strong enough to resist the backflow of blood when the reptile is positioned horizontally, not vertically. That makes me wonder if the more upright "Carolina Butcher" had stronger valves than modern crocodiles and alligators.

Humans and many other mammals have these valves, too, and for good reason. Giraffes, for instance, have extensive valves in their necks to prevent blood from draining away from their brains. But bipedalism strains these valves in us. With age, they can become leaky, causing blood to pool in the lower extremities. In humans, this can lead to varicose veins. The condition is especially common in women who have been pregnant, in part because carrying a child adds, on average, thirty-nine weeks of increased pressure to the circulatory system.

BIPEDALISM ALSO AFFECTS our sinus cavities, which fill with fluid, mucus, and all sorts of nastiness when we have an infection. The sinuses drain into the pharynx, which can be emptied with a simple clearing of the throat. Unfortunately for us, however, the duct emptying the maxillary sinus, located under the eyes, drains upward. That's why when you have a bad cold, you feel an uncomfortable pressure that, when severe, can even mimic the pain of a migraine headache.

Insight about this quirk of our anatomy comes from a study that compared goats and humans. Dr. Rebecca Ford, an ophthalmologist at King's College London, found that goats have no problems draining their maxillary sinuses. That's why many clinicians recommend that people with chronic maxillary sinusitis, an uncomfortable condition in which the mucus-clogged sinus becomes inflamed, get down on all fours like a standing goat. Humans haven't been bipedal long enough for evolution to come up with a better solution to this echo of our quadrupedal past.

PERHAPS THE MOST obvious side effect of bipedalism, however, is the toll it takes on our muscles and bones.

Quadrupedal backs are structured like suspension bridges with the guts hanging from a stable, horizontal string of vertebrae. But bipeds have taken the backbone and turned it ninety degrees. The human vertebral column is a series of twenty-four bones and discs stacked on top of one another. Paleoanthropologist Bruce Latimer of Case Western Reserve University imagines them as a stack of twenty-four cups and saucers precariously balancing much of the body's weight. To make matters worse, the stack isn't straight. It has three curves—an inward curve at the small of the back, an outward curve in the middle, and an inward curve of the neck vertebrae that brings the head over the shoulders.

There are advantages to these curves. Like springs, they help absorb compression forces during running, and they help move the bottom of the spine away from the birth canal. Because our spine has to bear the weight of the entire upper body, however, human vertebrae can, without any warning, break.

We are the only animals to have backbone fractures caused by nothing more than our own body weight, a risk that increases as we age. Perhaps not surprisingly, most of these fractures occur at weak points in the spine—the apexes of those curves. Annually, an estimated three-quarters of a million Americans suffer vertebral compression fractures.

But, that's not all. The weight of the body over the curved spine can shear the spines of the vertebrae (the parts you can feel if you rub your fingers over the middle of your back) right off the rest of the bone, causing one back bone to slide over the one below it. This condition, called spondylolisthesis, appears to be unique to humans and can pinch nerves and cause severe pain.

More common are slipped, or herniated, discs—rings of cartilage and gel-like material that provide padding between the vertebrae. The damage occurs when years of compression from walking upright causes the padding to spill past the bone and press against a nerve. The result is terrible, often debilitating, pain. In the lower back, herniated discs can press against the roots of the sciatic nerve and cause pain to shoot down the leg, a common condition known as sciatica.

As years of wear and tear further damage the discs, the padding between the vertebrae can completely deteriorate, making the bones rub against each other. This can cause osteoarthritis of the back, which in turn can lead to bone spurs that put pressure on spinal nerves and cause pain and weakness in the arms and legs. It is unusual to find bone spurs in the vertebrae of other animals. In adult humans, they are common. Stacking those twenty-four cups and saucers can also cause the spine to bend to the side, resulting in scoliosis, which develops in about 3 percent of school-aged children but is rare or entirely unheard of in other mammals.

IF YOUR BACK isn't bothering you yet, your knees might be act-
ing up. The human knee is not remarkably different from the knee
of any other mammal. It consists of two roundish knobs at the end
of our thigh bone (femur) that roll over a relatively flat top of the
shinbone (tibia). There's also a kneecap (patella) to help provide
leverage for our quadriceps muscles.

What is different about humans is that we put almost all of
our body weight directly on the knees rather than distributing it
across four limbs as quadrupeds do. When we walk, forces from
the ground travel up our legs like hammer strikes. At the knee,
these forces are surprisingly high. With each step, a force equivalent
to twice our body weight is absorbed by the knee. When we run,
the force is more than seven times our body weight. Some of these
forces are absorbed by the contraction of our muscles, and still others are dissipated by the cartilage padding between our bones. Over
time, these pads can deteriorate, and, because they don't have a
blood supply, they can't easily repair themselves. This wear and tear
eventually results in painful arthritis. In the United States alone,
more than 700,000 knees are replaced each year, in part because of
the damage bipedalism does to this joint.

The knee is vulnerable not only to gradual degradation but also
to sudden and severe injury. In 1951, the New York Yankees hosted
the New York Giants in game two of the World Series. When Giants legend Willie Mays slapped a fly ball to right center field,
Yankees center fielder Joe DiMaggio glided to his left and rookie
right fielder Mickey Mantle sprinted to his right to converge on it.
When Mantle saw DiMaggio settle under the ball, he pulled up
and caught his right cleat in a sprinkler system grate. Mantle's right
knee buckled, and he crumpled to the ground.

Mantle most likely tore his ACL (the anterior cruciate ligament), his MCL (the medial collateral ligament), and his medial meniscus—an injury orthopedic surgeons call the "unhappy triad." Mantle would go on to hit 536 career home runs in a storied Hall of Fame career, but his knee was never the same. Many believe this injury prevented him from being the best the game ever saw.

Soccer star Alex Morgan, Olympic skier Lindsey Vonn, former Patriots quarterback Tom Brady, and basketball pro Sue Bird have all torn their ACLs. It is a common injury, one that sidelines athletes for up to a year. For every sports star who tears the knee ligaments, there are tens of thousands of ordinary folks who do the same every year.

Functionally, the knee is simple. It flexes and extends. Anatomically, it's complex, held together with four ligaments that crisscross the joint and keep the femur attached to the bones of the lower leg. Two of the ligaments—the ACL, which crosses the front of the knee, and the PCL (the posterior cruciate ligament), which crosses the back—prevent the femur from sliding off the tibia. The other two—the MCL on the inside of the knee and the LCL (the lateral collateral ligament) on the outside—keep the knee from dislocating. These anatomical rubber bands, wonderful adaptations found across mammals, are subject to more strain in bipeds than in quadrupeds.

Close to 200,000 Americans blow out their ACLs every year. The injury is more common in women than in men. It is especially prevalent in sports such as basketball, soccer, field hockey, and football that require a lot of side-to-side movement. The high frequency of this injury is likely a result of moving on two, rather than four, legs, although the frequency in wild animals remains unknown.

Our knee ligaments are even more vulnerable because of adaptive changes to our pelvis and knees required for upright walking. Compared to apes, humans have wide hips and knocked knees that allow us to walk efficiently. This arrangement of our joints, however, results in the ends of our thigh bones meeting the knees at an angle. Forces traveling through angled objects tend to bend and shear those objects. Our knee ligaments are subject to much greater strains as a result. There are always trade-offs in evolution, and the knee provides a painful example of the costs of bipedalism.

IN 1976, WHEN twenty-one-year-old Van Phillips was a student at Arizona State University (ASU), he had a serious waterskiing accident that resulted in his left leg being amputated below the knee. He was given the standard prosthetic leg and sent home.

"I hated it," he told *OneLife Magazine* in 2010. "We've put man on the moon. And yet here I have this piece of crap. I knew, inside myself, that we could do a lot better."

He left ASU, became a student at the Northwestern University Prosthetic-Orthotic Center, and set to work on a better design inspired by cheetahs and pole vaulters. Years later, in 2012, the world watched in amazement as South African sprinter Oscar Pistorius used a prosthetic model based on Phillips's design to run the four-hundred-meter race at the Olympics in London.

Together, our two feet are composed of fifty-two individual bones—a quarter of all the bones in the human body. The bones are bound together by ligaments and held stiffly by the many muscles traversing the foot. In stark contrast, Phillips's prosthetic blade consists of a single moving element made of a compliant material stiff enough to drive the body forward but elastic enough to bend and recoil.

Unlike the blade, engineered in a lab, the human foot is the product of a long, complex, and nonlinear evolutionary history. However, one does not have to look far to see bladelike feet in the biological world. Large terrestrial birds such as ostriches and emus have feet that resemble Oscar Pistorius's prosthetic. Their ankle and foot bones are fused into a single rigid bone called the tarsometatarsus. They also have long, thick tendons that store elastic energy during bipedal locomotion, creating a recoil that puts a kick in their step. This anatomy allows ostriches to run almost forty-five miles per hour, twice the speed of a sprinting human.

No living mammals have joined us in our striding bipedal experiment, but if an asteroid hadn't wiped out the dinosaurs 66 million years ago (there is evidence that massive volcanic eruptions also contributed), scientists would be able to better analyze the convergent evolution of bipedalism. Many dinosaurs, including *T. rex*, were bipedal, and today's ostriches and emus trace their two-legged locomotion back to some of the earliest dinosaurs, which lived about 240 million years ago. This lineage has been bipedal about fifty times longer than our own.

Unlike us—the new kids on the bipedal block—terrestrial birds have fine-tuned their skeletons for this form of locomotion.

The earliest mammals living in the shadows of the bipedal dinosaurs were quadrupeds. Many lived in burrows, or in the forest canopies. One early skeletal modification in mammalian evolution was the development of the subtalar joint. This joint, located between the ankle bone (talus) and the heel bone (calcaneus) allows you to twist your foot inward and outward. These motions make a mammal's foot more mobile from side to side. The talus and calcaneus are fused in birds and are *next* to one another in both the reptilian ancestors of mammals and modern reptiles. But in the earliest mammals, the talus migrated to a position on top of the calcaneus, creating the new foot joint.

Try standing on one leg for a few seconds. Do you feel the wobble in your foot as you try not to topple over? Eventually, the contraction of muscles holding you upright becomes exhausting and you need a rest. Yet, flamingos can stand on one leg indefinitely without tiring. They don't wobble because they lack a subtalar joint. Their foot and ankle bones are fused together.

Humans have mobile ankles because we inherited our anatomy from ancestors who lived in the trees, where mobility was a significant advantage. For terrestrial bipeds, however, that mobility comes at a huge cost.

In the closing seconds of a 2013 game against the Atlanta Hawks, the late Los Angeles Lakers star Kobe Bryant dribbled to the right baseline and elevated to take a fallaway jump shot. His left foot landed awkwardly on the foot of his defender, and Bryant's rolled inward, pulling his ankle bones away from his leg. The anterior talofibular ligament that attaches the talus to the fibula overstretched, and Bryant gingerly walked off the court in excruciating pain.

At the 1996 Olympic Games in Atlanta, American gymnast Kerri Strug tore this ankle ligament. Moments later, with enough medical tape, adrenaline, and heart, she vaulted the American team to gold. Most people cannot walk, never mind vault, on a torn anterior talofibular ligament.

When anyone "sprains an ankle," it's usually the anterior talofibular ligament that overstretches and sometimes even tears. This band of tissue holding the talus to the fibula is the most often injured ligament in the human body. A million Americans sprain it every year, sometimes playing basketball but sometimes just stepping the wrong way on an uneven surface. It can take weeks to heal.

Because of bipedalism, human ankles are quite vulnerable to injury. To understand why, I traveled to Kibale Forest National Park in western Uganda.

ONE NEVER GETS dry in the rainforest. Even when it is not raining, the air is thick and wet. Sweat has nowhere to go but to soak clothes, hat rims, and socks. Elephant paths that cut through the dense forest are crisscrossed with vines that trip unsuspecting bipeds. Venomous snakes, large spiders, rash-causing plants, biting ants, and poacher traps abound. It is not a welcoming place for a New Englander, but it's where I had to go to study chimpanzees in their natural habitat.

The Ngogo chimpanzee community is 150 individuals strong and has been studied by John Mitani of the University of Michigan and David Watts of Yale University for two decades. I was there to see how the chimpanzees use their feet when they walk and climb. I didn't have to wait long to see what they could do.

On my first morning in the forest, Bartok, a large, stately alpha male, knuckle-walked to the base of a fruiting *Uvariopsis* tree, looked up to the crown with purpose, and climbed the foot-wide trunk as though he were walking up a flight of stairs. It appeared effortless. My eyes were fixed on his foot, and I couldn't believe what I was seeing. Bartok pressed the top of his foot against his shinbone and twisted his foot so that the bottom of it grasped the tree trunk. The first motion would have snapped my Achilles tendon. The second would have torn my anterior talofibular ligament.

For a month, I followed this group of chimpanzees and filmed nearly two hundred climbs. Each time, the chimpanzee positioned its foot in a way that would have caused severe tendon and ligament damage to most humans.

In humans, the Achilles tendon runs about halfway up the back of the lower leg to the base of the calf muscle. It is long and stores elastic energy to help put a kick in our step, particularly when we run. In chimpanzees, however, the Achilles tendon is only about an inch long. Most of the back of their leg is muscle, which is much more flexible than tendon, allowing the ankle more mobility as they climb. In other words, unlike us, chimpanzees don't have to worry about Achilles tendon ruptures.

They also don't have to worry about sprained ankles. They don't even have an anterior talofibular ligament.

The earliest apes were comfortable climbing trees, and not just because of their mobile foot joints. They also had a grasping big toe just as chimpanzees do today. Ape feet, the raw material from which the human foot evolved, were under intense natural selection pressure to be mobile, grasping appendages. Their foot muscles helped control the fine motions of their toes, important for grasping branches high in the forest canopy.

The human foot needs to be stiffer and more rigid to push off the ground during upright walking. Over the course of our evolutionary history, many moving parts of a once more mobile foot have been made more stable by ligaments, muscles, and a few subtle, bony alterations. These modifications, the biological equivalent of paper clips and duct tape, are wonderful examples of evolutionary tinkering.

Of course, the human foot does its job quite well. Natural selection has molded it into a structure that absorbs forces, stiffens during the propulsive phase of gait, and even has elastic structures including the arch and the Achilles. The intrinsic muscles, which previously controlled fine movements of our ancestors' grasping feet, now help support the arch. If these modifications had not evolved, it is likely that our ancestors would have been leopard food, and humans, as we know them, would not exist.

But because evolution makes small, gradual modifications to pre-existing structures, we have inherited a jerry-rigged solution for bipedalism that is effective enough to keep us on our feet but not elegant enough to do so without risk of pain and injury.

The plantar aponeurosis, for example, is a strong band of fibrous tissue that stretches across the bottom of the foot from the heel to the base of the toes. When stretched excessively it becomes inflamed, resulting in bone spurs and a painful condition called plantar fasciitis. Without this band, our foot would be too mobile to function properly, but it leaves us open to injury. We are also uniquely susceptible to collapsed arches, bunions, hammertoes, high ankle sprains, and all sorts of other maladies. It turns out that many of these foot ailments are exacerbated by the very technology that allowed humans to inhabit the globe: shoes.

Footwear helped humans spread into northern latitudes and eventually into the Americas. Today, shoes allow me to comfortably play blacktop basketball with my kids and hike in the woods after a nor'easter. High ankle boots protect against snakebites in the grasslands of Australia and sub-Saharan Africa. Footwear protects against broken glass at the beach or along city sidewalks. Or it simply allows one to buy something at the store since "no shirt, no shoes, no service." Without shoes, humans wouldn't have summited Mt. Everest or walked on the moon. They were and remain an important technological innovation. But as is true for many of our clever inventions, there are costs that accompany these benefits.

The bottom of your foot contains ten muscles, arranged in four layers. Some of these muscles maintain the foot's arch while others are critical for propelling us into our next steps. But most shoes, even the wholesome-sounding "arch-supporting" ones, can weaken these muscles. The result is a foot more prone to injury.

The Tarahumara are an indigenous people in Mexico known for their exceptional distance-running abilities. Their sandals are usually made from a piece of car tire rubber and held to the foot with string. Daniel Lieberman, a human evolutionary biologist at Harvard University and a distance runner himself, wondered about their feet. He traveled to the Sierra Tarahumara in northwest Mexico to study how they walk and run. He also used an ultrasound to measure the size of their foot muscles. Lieberman and postdoctoral researchers Nicholas Holowka and Ian Wallace reported in 2018 that the Tarahumara have higher arches, stiffer feet, and larger foot muscles than the typical American.

Maybe the Tarahumara are just genetically predisposed to have strong foot muscles? No. Elizabeth Miller of the University of Cincinnati's anthropology department worked with Lieberman's group and measured the size of two foot muscles in thirty-three runners. Half of the runners trained in their normal, cushioned running shoe. The other half transitioned slowly to a minimalist shoe—more like what the Tarahumara wear. After only twelve weeks, the minimalist shoe–wearers had increased the size of the two foot muscles by 20 percent and their arch was stiffer by a whopping 60 percent. Our feet change because of the shoes we wear, or don't wear.

Not only that, but without strong foot muscles, the plantar aponeurosis—the band of tissue that spans the bottom of the foot—can become overstrained, resulting in the stabbing pain of plantar fasciitis. Yet, as Harvard University biomechanist Irene Davis said, "We have lulled ourselves into thinking that our feet need cushioning to survive."

To boot, shoes no longer just protect feet. They are gendered symbols of social status, wealth, and power. Our feet pay the price. High-heeled shoes shorten the calf muscles and tighten the Achilles tendon, changing how we walk. Repeatedly squeezing the end of our foot into the narrow, pointed toe box of a shoe increases the chances of developing bunions and hammertoes. These damaging effects disproportionately impact women's feet and sometimes require surgical intervention.

"DR. HECHT HAS the best music," the operating-room nurse said to me.

I was dressed in a blue scrub suit, mask, and booties, a guest of Dr. Paul Hecht, an experienced orthopedic foot and ankle surgeon. A man in his forties lay on the operating table at the Dartmouth-Hitchcock Medical Center. The winter before, he had slipped on ice and fractured his right ankle. Screws had been inserted into the bone to help with the healing, but this hadn't worked well. He required ankle fusion surgery.

Stevie Wonder sang "Don't you worry 'bout a thing" through the overhead speakers.

The first cuts were delicate as Dr. Hecht meticulously separated the superficial tissues—skin, subcutaneous fat, and muscle—to reach the ankle joint.

Then, the scene began to look more like Home Depot than an operating room. The drills came out first to remove the old screws from the tibia.

"You know what a lag screw is?" Dr. Hecht asked me.

"Um . . . yeah," I said. I'd used them to anchor my kids' treehouse to a large oak in our backyard. I didn't expect such large, multi-threaded steel screws in an operating room.

An electrocautery pen was used to make fine cuts and to stop bleeding from small veins that inevitably get nicked in such a surgery. The smell of singed tissue filled the room. The talus bone was cranked away from the shin like a car being jacked to replace a flat.

With the joint exposed, things got messy. Cartilage was scraped from the joint with a tool that looked like a melon baller. Dr. Hecht then used an electric drill to bore holes in the joint and encourage bleeding that would bring bone-building cells to the area and start fusing the joint. As the drill buzzed, flakes of bone took flight, so I took a few steps back. Then, with a hammer and chisel, Dr. Hecht elevated the outer layer of bone into small chunks that resembled fish scales to increase the surface area over which healing can occur. Finally, a soupy blend of live bone cells and microscopic bony scaffolding that looked like slime from *Ghostbusters* was grafted onto the broken bones to accelerate healing and the growth of new bone. Before Dr. Hecht became an orthopedic surgeon, he was trained as a woodworker. It made sense.

Later that day, I watched as the back of a middle-aged woman's heel was sheared away with an electric saw to remove painful bone spurs. In another patient, an electric drill rounded off the joint of the big toe to fix arthritis.

Orthopedic surgery is a multibillion-dollar industry, and it has our evolutionary history to thank for its success.

To be sure, some of our foot maladies are the result of our sedentary lifestyle and our decision to wear shoes. But foot pathologies are common in hominin fossils from long before the invention of footwear. The negative consequences of upright walking have been with us for a long time.

It turns out that these pathological old bones reveal something else about being human—something that helps us return to the start to unravel the mystery of how an ape first began moving on two legs.

CONCLUSION:
THE EMPATHETIC APE

What a frail, easily hurt, rather pathetic thing a human body
is, naked; somehow a little unfinished, incomplete!

—*D. H. Lawrence,* Lady Chatterley's Lover, *1928*

Bipedalism set in motion all of the major evolutionary events in the human lineage, from tool use and cooperative parenting to trade networks and language, eventually allowing us, a once humble ape standing in the Miocene forests, to populate the globe.

But it is still a wonder we are here. We are pathetically slow, at best a third of the speed of a typical galloping quadruped of our size. The two leopard fang holes in the back of the head of an *Australopithecus* fossil are a gruesome reminder that there were evolutionary consequences for our lack of speed. We are unstable on our two legs, accidental falls accounting for over half a million deaths around the world each year. The short, squat pelvis that biomechanically adapts humans for efficient bipedal travel also forces a baby to corkscrew through the birth canal during delivery, making childbirth difficult and sometimes dangerous. After birth, our adventurous toddlers fearlessly and foolishly wobble right off gaps in an experimental runway if not supervised. And as we age, bipedalism takes its painful toll on our backs, knees, and feet.

The advantages of bipedal locomotion obviously outweigh the

costs. Otherwise, we would have gone extinct long ago. But given the many downsides to upright walking and how rare this form of locomotion is in the animal world, I've wondered what tipped the scales toward survival rather than extinction.

The answer may be found with one of the most wonderful and mysterious aspects of the human condition. To understand, we have to revisit the human fossil record.

SOME FOSSILS HAVE names like "Lucy" or "Sue." Most have names like KNM-ER 2596.

"KNM" stands for "Kenya National Museum," the current location of this particular fossil. "ER" represents "East Rudolf," indicating that the fossil was found along the eastern shore of Lake Rudolf, the colonial name for Lake Turkana in northern Kenya. The number 2596 means that it was the 2,596th fossil discovered at that locality, its recovery made in 1974. Since then, more fossils have been collected in that area, bringing the current tally to almost 70,000.

KNM-ER 2596 is a small, cracked piece of distal tibia, the scientific name for the bottom of the shinbone. Where it once met the ankle joint, it is expanded and filled with spongy bone, a clear indication that this fossil belonged to an upright walking hominin.

From the size of the bone, we can estimate that this individual weighed a little less than seventy pounds, which is about the size of Lucy. A faint line around the perimeter of the bone is a sealed growth plate, showing that this hominin had reached full size shortly before death. Taken together, these clues suggest a female in her late teen years. She died about 1.9 million years ago, according to the amount of radioactivity present in the ash layers that surrounded the fossil. Several carnivore tooth impressions reveal the probable cause of death.

We aren't sure what species KNM-ER 2596 belonged to because several different kinds of hominins lived at the time. But something is off about this bone. It doesn't look exactly like the shinbone from Lucy's skeleton or from any other bipedal hominin. The medial malleolus, the roundish knob on the inside of the ankle, is unusually small and atrophied. The ankle joint is angled in a peculiar manner. These odd anatomies sometimes are found today in people who break their ankles in childhood and never have the bones set properly.

There were no doctors or hospitals 1.9 million years ago, of course, but after this little hominin broke her ankle, leaving her helpless in a world of predators, she didn't die. Not then. She lived long enough to heal and grow up.

Fossils are just rocks, but they tell extraordinary stories. Imagine the scene along the eastern side of Lake Turkana 1.9 million years ago. The sun rose, casting golden light across the sprawling grasslands. In a gallery forest hugging a nearby river, monkeys awoke with a racket. Ancestors of zebras, antelopes, and elephants munched their breakfasts, occasionally lifting their heads to scan for predators lurking in the tall grass.

From the safety of their trees, hominins watched the scene unfold. They didn't dare come to the ground. The predators were hungry, and hominin was on the menu. But once the sun rose high enough to drive the large cats into the shade, the hominins climbed down to search for food. They gathered grubs, tubers, fruit, seeds, immature leaves, and maybe even meat clinging to the bones of kills the cats made during the night.

One of those hominins was KNM-ER 2596. She was with her family and friends, a group of perhaps two or three dozen. Her mother wasn't feeding her anymore since there was another baby to take care of, but KNM-ER 2596 helped carry the baby as they foraged.

As the sun set, she retreated back into a tree and made a nest for the night. Perhaps she looked up and wondered about the points of light in the sky.

One day, KNM-ER 2596's life changed dramatically. Maybe she fell out of a tree. Maybe she stumbled into a ditch. However it happened, her ankle twisted, the ligaments tore, and the bone shattered. She sprawled on the ground crying in pain, crying for help. Her mother ran over to help but couldn't put her baby down—not in the open grassland with predators nearby. The rest of the group approached, worried looks on their faces, knowing that the commotion would soon attract large cats and hyenas.

The safest thing for the group was to abandon her there, but that is not what happened.

Perhaps some of them carried her to a wooded area and helped her get into a tree. Perhaps the tree was fruiting, and she could nibble without leaving the safety of its branches. Perhaps others brought her grubs, a chunk of antelope, or a handful of seeds. Perhaps it was the rainy season, and she could lick water off the leaves.

If we had found more of her skeleton, more of her story would be revealed, but a precious scrap of her shinbone is the only evidence we have that she existed. Do we *know* that other members of the group cared for her while she healed? No, but it is difficult to imagine how she survived otherwise. KNM-ER 2596 slowly got better, but she never lost her limp.

When a quadruped, such as a zebra or an antelope, is badly injured, it hobbles, but it can still walk. When a biped is severely injured, it can no longer walk. Bipedalism not only leaves us vulnerable to leg and foot injuries but makes us particularly feeble when they happen.

If KNM-ER 2596 were the only example of a hominin surviving a catastrophic injury, we'd note how lucky she was and put her in a footnote. But she was not the only one who needed help to survive injury or disease. There were others—many others.

A 3.4-million-year-old *Australopithecus afarensis* skeleton discovered at the Woranso-Mille, Ethiopia, site by Yohannes Haile-Selassie had a healed ankle fracture just like KNM-ER 2596's. Around the same time KNM-ER 2596 was recovering on the shores of Lake Turkana, a hominin known as KNM-ER 738 broke her left femur. It was a spiral fracture, the kind today's emergency-room doctors often see after car crashes and skiing accidents. Typically, six weeks of complete immobilization are needed before the patient can walk again. KNM-ER 738 should not have survived. But this fossil, discovered by Richard Leakey's team in 1970, has a thickened area of bone called a callus, evidence that she healed and lived on.

Rings of inflamed bone circled and thickened the skeleton of a *Homo erectus* whose fossil is named KNM-ER 1808. At first, scientists blamed this on an overdose of vitamin A, a condition that afflicted early-twentieth-century shipwrecked sailors who ate too much seal liver, developed similar growths, and died. Others propose that the condition was caused by yaws, a bacterial infection that is rarely fatal but disfigures people today. No matter the cause, KNM-ER 1808's bone inflammation would have been painful and debilitating. But this *Homo erectus* kept eating, moving, and breathing. It is hard to imagine he could have done so without help.

And the list goes on. The 1.49-million-year-old Nariokotome *Homo erectus* child looks like he had scoliosis. A 1.8-million-year-old partial foot fossil from Olduvai Gorge, Tanzania, has bony growths indicating severe arthritis. Hominin leg bones found nearby exhibit diseased bone from a severe high ankle sprain. Vertebrae from 2.5-million-year-old cave sediments in South Africa contain rings of bone consistent with serious lower back arthritis. In the same cave deposits, researchers discovered a healed compression fracture in the ankle of an *Australopithecus*. Karabo, the *Australopithecus sediba* skeleton found by nine-year-old Matthew Berger and his dog Tau, had a tumor on his vertebrae that would have throbbed and ached. In each case, these individuals would have benefited from the assistance of others.

Life was hard for our ancestors. Moving on two legs made it harder. Every day, they competed for food with other hominin species while trying to avoid fearsome predators. With all of these threats, they needed to fiercely defend themselves against the dangerous "other" while simultaneously directing empathy toward their own.

Harvard University primatologist Richard Wrangham calls this the "Goodness Paradox." How can we humans be both cruel and compassionate? Scholars have debated the essence of human nature for centuries. Are we innately violent and restrain our aggressive tendencies through rules and group norms, or are we peaceful by nature and become aggressive in oppressive societies that celebrate violence and the patriarchy?

All mammals, including humans, are behaviorally flexible. They can be nurturing one moment and violent the next. Adorable otters hold hands and lovingly groom one another, but they attack and forcibly copulate with baby seals. Elephants nurture newborns one moment and trample a human safari-goer the next. The domestic dog is a member of the family in more than 50 million U.S. homes. Dogs fetch, nuzzle, and lick, but they also bite. Our furry friends bite 4.5 million Americans a year, resulting in 10,000 hospital visits and, in 2019, 46 deaths.

Mammal behavior is a dance of hostility and harmony.

Our closest cousins, chimpanzees and bonobos, are often considered behavioral opposites. At times, chimpanzees are ruthless killers, while bonobos generally are free-spirited pacifists. Those who see humans as naturally violent often cite research on chimpanzees to support their argument. Those who regard humans as peaceful by nature cite bonobos. Reality is more nuanced than that.

In Uganda's Kibale Forest National Park in 2006, I watched Miles, one of the high-ranking males in his group, viciously beat a female chimpanzee. Her desperate efforts to escape repeatedly failed as he grabbed her by the leg, dragged her back, and beat her with closed fists. Two days earlier, however, I watched a tranquil Miles lying on his side, playing with a youngster. He was gentle, loving.

A year later, I accompanied the same group of chimpanzees on patrol. A dozen males knuckle-walked with purpose to the boundary of their territory. They sniffed the air, sometimes standing bipedally to listen and look for the enemy. They moved in unsettling silence. That day ended uneventfully, but a week before I arrived, the same group encountered a chimpanzee from a neighboring group and beat him to death.

Bonobos, however, have never been seen engaging in territorial killings. When they meet up with neighbors, they groom them, share food, and even have sex. In their resource-rich forest, the best behavioral strategy appears to be to make love, not war, but that does not mean bonobos are pacifists. They hunt and eat meat, and squabbles between group members in their female-dominated societies occasionally turn violent. Sometimes, female bonobos form coalitions to attack and subdue aggressive males.

"THE POTENTIAL FOR good and evil occurs in every individual," Richard Wrangham wrote in *The Goodness Paradox*. Does the fossil record provide any insights into the balance between aggression and amity in the human lineage?

In northern Spain's Atapuerca Mountains, a team of paleoan-thropologists has recovered 7,000 hominin fossils from a half-million-year-old cave called Sima de los Huesos, or "pit of the bones." They have sorted the jumbled remains into twenty-eight partial skeletons. The fossils are the oldest to still preserve DNA, which has revealed that the Atapuerca people were ancestors of Neandertals.

One of them, nicknamed "Benjamina" by the researchers, was a child who died when she was about seven years old. Her deformed skull indicates that she had a severe case of craniosynostosis, a condition that causes mental impairment. Caring for a child for seven years takes devoted caretakers anyway, but Benjamina's case would have required care above and beyond. On the other hand, another individual whose bones came to rest not far from Benjamina's offers evidence of brutality. He had been beaten to death with a rock. Two blows to the forehead, just above the left eye, penetrated the skull and exposed his brain. His body was tossed down a natural sink-hole and into the same pit of bones.

Thirty-six thousand years ago, someone took a sharp rock—perhaps a hand ax—and slashed the top of a Neandertal's head near modern-day Saint-Césaire, France. But his fossil shows healed bone around the wound, evidence that he continued to live.

One hundred and fifty thousand years ago, a young girl living in Lazaret Cave near modern Nice, France, suffered a blow to the right side of her head. Maybe she was fooling around and fell. Perhaps a friend lobbed a rock that hit her by accident. Maybe a member of her group—or of a neighboring clan—smashed her head deliberately. No matter how it happened, the fossil she left behind shows that she was badly hurt. Even though she must have bled profusely suffering that kind of head wound, she healed. Someone must have tended to her injuries until she recovered.

In 2011, Xiu-Jie Wu of the Chinese Academy of Sciences published the results of her analysis of a nearly 300,000-year-old skull from southern China. He, too, suffered a blow to the top of the head and healed. Wu and her colleagues went on to document additional instances of traumatic violence—more than forty head wounds—in fossils of our ancestors. But in almost every case, the victim survived and healed, and most could not have done so without help.

HUMANS ARE CLANNISH. Much like chimpanzees, we often limit our altruism to those perceived as part of our group. We can unleash terrible violence on those we define as "other," sometimes to seize wealth or territory but often merely because they worship a different god, have a different skin color, speak a different language, or live under a different flag. Yes, humans excel at cooperating with one another, but one of the things we cooperate best at is killing large numbers of other human beings.

From the club-wielding apes of *2001: A Space Odyssey* to the false but still-pervasive "Man the Hunter" notion that our evolution was driven in part by hunger for the meat of large animals, our undeniably violent and aggressive tendencies have dominated the narratives we have constructed about our past. And yet, our evolutionary journey has also equipped us with an extraordinary capacity for empathy. Too often, we've brushed aside the better angels of our nature and ignored the fact that, as with Wu's forty hominins who survived head wounds because they had help, conflict and empathy are linked.

"Your heart does roughly the same thing whether you are in a murderous rage or having an orgasm," Stanford University neurobiologist and primatologist Robert Sapolsky wrote in his book *Behave: The Biology of Humans at Our Best and Worst*. "The opposite of love is not hate, it's indifference." But indifference is not something I see in the human fossil record.

Recall the 3.66-million-year-old Laetoli footprints. The smallest individual appears to have walked with a severe limp, her foot angled almost thirty degrees from her direction of travel. But she didn't walk alone. She was with others—her helpers.

Lucy must have had helpers, too. Her femur contains a sharp arc of infected bone where her hip muscles would have been attached. Perhaps it was caused by a thorn that stuck deep in her side. Perhaps her tendon tore from the bone as she desperately escaped from the jaws of a predator. She got away, but her hip would have hurt and caused her to limp.

Lucy also had back problems. Even though she was young, four of her backbones had developed strange growths similar to those found in people today with a skeletal disorder called Scheuermann's disease. This condition may have given her a hunchback and compromised her ability to walk. Life was hard and painful for our science's icon.

Even more telling is the way Lucy's species gave birth.

In 2017, I worked with anthropologists Natalie Laudicina, Karen Rosenberg, and Wenda Trevathan to reconstruct how they did it. From the shape of her pelvis, we determined that it was impossible for an *Australopithecus* to be born facing forward as happens in most ape births. Instead, the baby rotated as it entered the birth canal. As the baby reached the midplane, it had to keep rotating to get its shoulders through. Although our simulations did not require a full 180-degree rotation, the baby still had to be born facing backward, in the occiput anterior position, as most humans are born today. It would have been dangerous for Lucy's species to give birth alone.

To paleoanthropologists, this means Lucy had helpers. Midwifery must go back at least 3.2 million years to the time of *Aus-*

tralopithecus. As Rosenberg wrote, "Midwifery . . . is the 'oldest profession.'"

Chimpanzees, whose spacious pelvises don't require their babies to rotate through their birth canals, typically give birth alone, but in bonobos, our other first cousins with roomy pelvises, birth is not always a solitary experience.

In 2018, Elisa Demuru, a postdoctoral researcher at the University of Lyon in France, published observations of three captive bonobo births. Other females were present and even assisted by holding the infant as it was being born. A few years earlier, Pamela Heidi Douglas, a scientist at the Max Planck Institute for Evolutionary Anthropology in Leipzig, Germany, observed a rare daytime birth of a wild bonobo in the forests of the Democratic Republic of the Congo. Again, other females were present.

In this three-pronged family of humans, chimpanzees, and bonobos, chimpanzees are the odd apes out. Perhaps they changed their birth behavior from a social to a solitary one over the course of their evolutionary history.

It seems probable, then, that when the last common ancestor of humans, chimpanzees, and bonobos gave birth, other females were present and ready to assist. Perhaps social support during bipedal hominin births *predates* the physical need to have helpers. Perhaps rotational birth, made necessary by pelvic changes that accompanied bipedalism, was possible only because female assistance was already part of the behavioral repertoire of our hominin ancestors.

In the chicken-and-egg scenario of birth assistance and rotational birth, the logical conclusion is that helpers came first.

UPRIGHT WALKING IS intimately linked to our evolution as a social species. The evidence suggests that our bipedal ancestors not only assisted at births but also looked after infants as their mothers foraged. They formed communities that kept their young safe as their brains grew and they learned the ways of their group. Too slow to flee and too small to ward off attacks alone, they had to look out for one another to survive.

Today, we take for granted this ancient foundation of trust, generosity, and cooperation even as our children fearlessly stumble into their first steps, confident that a caretaker is nearby, buffering them from danger. We subconsciously coordinate our strides with those around us, as we have done for millennia.

Bipedalism evolved in conjunction with empathy and drove the development of technology. Along with intelligence, it ultimately resulted in modern medicine, hospitals, wheelchairs, and prosthetics. The evolution of able-bodied walking in a social, empathetic ape is what made *not walking* for nearly 3 million disabled Americans possible.

Primatologist Frans de Waal has written that empathy starts with the "synchronization of bodies." By walking in stride with those around us, we cannot help but put ourselves in one another's shoes.

Like many ideas, the link between bipedalism and social tendencies traces back to Darwin. In 1871, he wrote:

> *In regard to bodily size or strength, we do not know whether man is descended from some small species, like the chimpanzee, or from one as powerful as the gorilla; and, therefore, we cannot say whether man has become larger and stronger, or smaller and weaker, than his ancestors. We should, however, bear in mind that an animal possessing great size, strength, and ferocity, and which, like the gorilla, could defend itself from all enemies, would not perhaps have become social; and this would most effectually have checked the acquirement of the higher mental qualities, such as sympathy and the love of his*

fellows. Hence it might have been an immense advantage to man to have sprung from some comparatively weak creature.

Although his overall point is a good one, there are factual errors in this passage. Chimpanzees are not small and weak; they are very strong. Gorillas are less ferocious and more social than Darwin describes. And it is a mistake to assume that a caring, social species is "weak."

The notorious mobster Al Capone may have said, "Don't mistake my kindness for weakness." An almost identical statement—"Don't ever mistake . . . my kindness for weakness"—is attributed to the Dalai Lama. This nicely illustrates our remarkable behavioral flexibility. We are peaceful and violent, cooperative and selfish, empathetic and apathetic. De Waal wrote, "We walk on two legs: a social and a selfish one."

We tend to shine a spotlight on our selfish tendencies and take our sociality for granted. Every day, millions of generous, thoughtful, kind, life-changing acts are carried out by humans without much notice. But when we stray from the cooperative side of our nature and commit acts of greed and violence, it is aberrant enough to be newsworthy.

Bombarded in our twenty-four-hour news cycle with examples of human cruelty, we often overlook how remarkably cooperative and tolerant we can be. Helping one another comes naturally to us: holding a door open for a neighbor, donating spare change to a panhandler, passing a plate to share food with others. These are such everyday occurrences that human kindness, like walking, has become pedestrian.

It is worth noting that humans, and our hominin ancestors, are by no means the only creatures that cooperate or demonstrate empathy. These behaviors for maintaining social cohesion have been widely observed in the animal kingdom. For example, ants and bees cooperate far more fully and efficiently than we do. Empathy has been observed in species as varied as elephants, dolphins, and dogs.

And echoes of *our* compassionate nature appear in our ape cousins.

In 1974, Penny, a three-year-old chimpanzee, fell into water surrounding her island enclosure at the Oklahoma Institute of Primate Studies and began to drown. Washoe, an unrelated nine-year-old female chimpanzee, leapt an electrified fence and pulled her to safety. In 1996, Binti-Jua, a female western lowland gorilla, scooped up and cradled a three-year-old human boy who had fallen into her enclosure at the Brookfield Zoo outside Chicago before carrying him to safety. In early 2020, an orangutan was photographed extending a helping hand to a man up to his waist in water. Bonobos, the most empathetic and altruistic of the great apes, routinely share food, even with strangers.

WHAT IT TOOK for the seeds of cooperation and altruism to burst forth in the human lineage were the colossal challenges created by upright walking.

In 2011, paleoanthropologists Don Johanson and Richard Leakey joined neurosurgeon and medical journalist Sanjay Gupta for a public event at the American Museum of Natural History in New York. The last time the paleoanthropologists had shared a stage in 1980, Leakey stormed off in a rage over their differing interpretations of old bones they were pulling from ancient African sediments. But a few decades later, these two silverbacks of our field had mellowed enough to jointly reflect on their careers.

During the Q&A, Gupta asked what made us human. Leakey spoke first about how he had lost his legs when they were crushed in an airplane crash in 1993, and how he now walked on artificial limbs. He said:

> *If you are a two-legged creature and you have no legs, you don't get very far . . . being a uniped is no better than being a no-ped. Whereas, if you're a chimpanzee, or a baboon, or a lion, or a dog, and you have four legs, you can lose one and do perfectly well. Now, once we became bipedal . . . bonding and social interactions take on a totally different, not just meaning, but value. And I do not believe bipedal primates could have survived unless they had, in addition to being bipedal, changed the way they think in terms of altruism and in terms of social networking and social connections.*

It could be, then, that one of the most mysterious aspects of the human condition—our capacity for selflessness—arose out of our vulnerabilities as bipeds in a dangerous world. Yes, our survival was, and for many continues to be, a struggle, but as descendants of bipedal hominins, our evolutionary journey continues because empathy, cooperation, and generosity evolved in lockstep with our distinctive form of locomotion.

I would argue that the human experiment would not have been possible unless we descended from social apes capable of empathy—that bipedalism could have evolved only from a lineage that had developed the capacity for tolerance, cooperation, and caring for one another. Bipedalism in an overly aggressive ape with purely selfish tendencies and a low tolerance for other group members would have been a recipe for extinction.

For the film *Contact*, Carl Sagan wrote of humans: "You're an interesting species. An interesting mix. You're capable of such beautiful dreams, and such horrible nightmares. You feel so lost, so alone, only you're not. See, in all our searching, the only thing we've found that makes the emptiness bearable, is each other."

After millions of years and dozens of evolutionary experiments, we humans are the last bipedal ape on Earth. As we stride forward as a species into uncertain and unsettling times, it helps to glance back over our shoulder at the trail we've left. We've traveled far, and overcome much, together.

It is time to embrace the lessons the bones of our ancestors teach us and construct a new human origin story in which the evolutionary success of this extraordinary upright ape is attributed in large part to our capacity for empathy, tolerance, and cooperation.

.

ACKNOWLEDGMENTS

This book would not have been written without the support of my favorite bipeds in the world. Ben and Josie: thank you for your patience, humor, love, and advice as your dad was busy writing his book. Follow your own path, but always be there for one another. And may the steps you take lead to happiness and to a more just world. And to Erin—you have always believed in me and lifted me up. I cannot imagine a better partner to walk through this life with than you.

I am fortunate to have a loving family who encouraged me through this process. Thank you, Rich, Mel, Deana, Chris, Mom, Aunts Ginny and Mary, Kittie, Dadou, Patricia, Mikaila, Mike, Lorrie, Adam, Ashley, Alex, Lillian, Jake, Ella, Anthony, Ian, Jameson, and Wyatt. And thanks to my favorite quadruped, Luna, for taking long walks with me as I struggled to write.

When I was in sixth grade, I got into trouble with my teacher, and my punishment was to transcribe my textbook word for word onto yellow, lined paper. My father generally agreed that misbehaving children should be disciplined. But he lost his mind when I told him how I had been sentenced. He contacted the school and demanded that I receive a different punishment. To my father, writing was not a penalty. It was a gift. He is right. Thank you, Dad, for reading every line of this book—several times—and for helping me find my voice with your many helpful edits. Talking writing and science with you was one of my favorite parts of writing this book.

My agent Esmond Harmsworth at Aevitas believed in this book long before I did. Thank you for meeting me for lunch at BU and for your guidance and wisdom. The team at Aevitas, Chelsey Heller, Erin Files, Sarah Levitt, Shenel Ekici-Moling, and Maggie Cooper, are extraordinarily good at their jobs. It has been a pleasure working with them.

Thank you to Gail Winston, my skilled and brilliant editor at HarperCollins, Alicia Tan, Sarah Haugen, Becca Putman, Nicholas Davies, and to the entire team at HarperCollins for making this an enjoyable experience at every stage in the process. I hope this is just the beginning. I am grateful to Fred Wiemer for his skillful and careful copyediting.

Everything I am and everything I do as a scientist and a science communicator is because of Lucy Kirshner and Laura MacLatchy. Lucy, you are at the meeting point of a Venn diagram that includes science, science literacy, museum education, Laetoli, Africa, Ann Arbor, Acton, and so many other places and ideas that have shaped who I am. Laura, I could not asked for a better advisor in science and in life. Thank you for taking a chance on me in 2003 and for your continued guidance and friendship.

I am grateful for the many scientists, writers, teachers, and scholars who graciously took the time to talk with me about their work: Karen Adolph, Zeray Alemseged, Hutan Ashrafian, Kay Behrensmeyer, Riley Black, Madelaine Böhme, Greg Bratman, Michel Brunet, Chris Campisano, Susana Carvalho, Rama Chellappa, Habiba Chirchir, Zach Cofran, Omar Costilla-Reyes, Elisa Demuru, Todd Disotell, Holly Dunsworth, Kirk Erickson, Dean Falk, Simone Gill, Yohannes Haile-Selassie, Carina Hahn, Shaun Halovic, Will Harcourt-Smith, Sonya Harmand, Katerina Harvati, Paul Hecht, Amanda Henry, Kim Hill, Ken Holt, Jonathan Hurst, Christine Janis, Stephen King, John Kingston, Bruce Latimer, I-Min Lee, Sally

Le Page, Dan Lieberman, Paige Madison, Antonia Malchik, Ellie McNutt, Anne McTiernan, Fredrick Manthi, Stephanie Melillo, Joann Montepare, Steven Moore, W. Scott Pearson, Bente Klarlund Pedersen, Martin Pickford, Herman Pontzer, Stephany Potze, Lydia Pyne, Dave Raichlen, Phil Ridges, Tim Ryan, Brigitte Senut, Liza Shapiro, Sandra Shefelbine, Scott Simpson, Tanya Smith, Michael Stern, Ian Tattersall, Randall Thompson, Erik Trinkaus, Peg van Andel, Michelle Voss, Cara Wall-Scheffler, Carol Ward, Anna Warrener, Jacqueline Wernimont, Jennifer Weuve, Katherine Whitcome, Bernard Wood, Lindsay Zanno, Bern Zipfel, and Ari Zivotofsky. I apologize if I inadvertently forgot anyone.

A special thank you to my colleagues who opened their labs, field sites, operating rooms, and zoos to me: Karen Adolph, Madelaine Böhme, Omar Costilla-Reyes, Todd Disotell, Paul Hecht, Jonathan Hurst, Nathaniel Kitchel, Charles Musiba, Martin Pickford, Phil Ridges, Michael Stern, Cara Wall-Scheffler, and Lindsay Zanno. I'm especially grateful to my colleagues Bern Zipfel, Lee Berger, Charles Musiba, and Yohannes Haile-Selassie: your work inspires me and your friendship means even more. Thanks to my friends, family, and colleagues who read large sections of the book and helped improve its accuracy and readability: Nathaniel Kitchel, Simone Gill, Karen Adolph, Dave Raichlen, Brian Hare, Scott Simpson, Blaine Maley, Shirley Rubin, Melanie DeSilva, Paul Hecht, Adam van Arsdale, Cara Wall-Scheffler, and Lindsay Zanno.

I am fortunate to have supportive, brilliant, thoughtful colleagues in the anthropology department at Dartmouth College. Thanks especially to Nate Dominy and Zane Thayer for always asking the right questions and for inspiring me with your unrelenting curiosity about the world. This book took its initial shape thanks to the help of the talented Learning Design team at Dartmouth and the development of the MOOC Bipedalism: The Science of Upright

Walking. Special thanks to Adam Nemeroff, Sawyer Broadley, Josh Kim, and Mike Goudzwaard.

I am particularly thankful for my students and for the observations and questions that constantly keep me on my toes. There are too many to list, but many ideas in this book derive from conversations I've had with the students I have taught at Worcester State University, Boston University, and Dartmouth College. My former and current graduate students and undergraduate researchers are forever challenging my ideas with fresh eyes and brilliant insights. Many thanks to Ellie McNutt, Kate Miller, Luke Fannin, Anjali Prabhat, Sharon Kuo, Eve Boyle, Zane Swanson, Corey Gill, Jeanelle Uy, and Amey Y. Zhang.

Finally, a special thank-you to Alex Claxton. In a book that includes archosaurs, hominins, early mammals, and *Andrewsarchus*, there is no one better skilled or informed to fact-check these pages. I am awed by your breadth of knowledge and unending curiosity. I can't wait to read *your* first book.

Despite the many efforts I took to present as accurate a picture of our current understanding of bipedal evolution and the many downstream effects of humans moving on two legs, mistakes have undoubtedly crept into these pages. Any and all errors are my own.

NOTES

INTRODUCTION

xv   "There's an old story about a centipede": Duncan Minshull, *The Vintage Book of Walking* (London: Vintage, 2000), 1.

xv   Of the 636 taken: "New Jersey Division of Fish & Wildlife," last modified October 10, 2017, https://www.njfishandwildlife.com/bearseas16_harvest.htm.

xv   there was outrage: Daniel Bates, "EXCLUSIVE: Hunter Who Shot Pedals the Walking Bear with Crossbow Bolt to the Chest Is Given Anonymity over Death Threats," *Daily Mail*, November 3, 2016, https://www.dailymail .co.uk/news/article-3898930/Hunter-shot-Pedals-bear-crossbow-bolt-chest -boasting-three-year-mission-given-anonymity-death-threats.html.

xvi   One has over a million views: "Pedals Bipedal Bear Sighting," last modified June 22, 2016, https://www.youtube.com/watch?v=Mk-HHyGRSRw.

xvi   another over 4 million: "New Jersey's Walking Bear Mystery Solved," August 8, 2014, https://www.youtube.com/watch?v=kcIkQaLJ9r8&t=3s.

xvi   chimpanzees giving hugs: See Frans de Waal, *Mama's Last Hug: Animal Emotions and What They Tell Us About Ourselves* (New York: W. W. Norton, 2019). Video of encounter: https://www.youtube.com/watch?v=INa-oOAexno.

xvi   In 2011, news spread that a male silverback: "Gorilla Walks Upright," CBS, January 28, 2011, https://www.youtube.com/watch?v=B3nhz0FBHXs. "Gorilla Strolls on Hind Legs," NBC, January 27, 2011, http://www.nbcnews .com/id/41292533/ns/technology_and_science-science/t/gorilla-strolls -hind-legs/#.XllgdpNKhQI. "Walking Gorilla Is a YouTube Hit," BBC News, January 27, 2011, https://www.bbc.co.uk/news/uk-england-12303651.

xvi   Upright-walking-gorilla mania: "Strange Sight: Gorilla Named Louis Walks like a Human at Philadelphia Zoo," CBS News, March 18, 2018, https:// www.youtube.com/watch?v=TD25aORZjmc. I visited Ambam in February 2019 and Louis in October of that year. Their keepers were very helpful and knowledgeable about the gorillas, and I had a marvelous time watching these magnificent cousins of ours. For the several morning hours I observed them, both gorillas knuckle-walked from one spot in their enclosure to another. I never saw them walk bipedally. Even those individual apes who are more comfortable moving on two legs still only do it occasionally.

xvi   Faith the dog: "Things You Didn't Know a Dog Could Do on Two Legs," Oprah.com, https://www.oprah.com/spirit/faith-the-walking-dog-video.

xvii   video of a bipedal octopus: "Bipedal Walking Octopus," January 28, 2007, https://www.youtube.com/watch?v=E1iWzYMYyGE.

PART I: THE ORIGIN OF UPRIGHT WALKING

1   "All other animals look downward": Ovid, *Metamorphoses, Book One*, trans. Rolfe Humphries (Bloomington: Indiana University Press, 1955).

CHAPTER 1: HOW WE WALK

3   "Walking is falling forward": Paul Salopek, "To Walk the World: Part One," December 2013, https://www.nationalgeographic.com/magazine/2013/12/out-of-eden.

3   Even Plato recognized: From Diogenes Laërtius, *The Lives and Opinions of Eminent Philosophers*, trans. C. D. Yonge (London: G. Bell & Sons, 1915), 231.

4   Bipedalism has since made its way into our words: I've discovered that this trope of listing off names and metaphors for walking is a common practice. Variants on it are in Rebecca Solnit, *Wanderlust: A History of Walking* (New York: Penguin Books, 2000); Antonia Malchik, *A Walking Life* (New York: Da Capo Press, 2019), 4; Geoff Nicholson, *Lost Art of Walking* (New York: Riverhead Books, 2008), 17, 21–22; Joseph Amato, *On Foot: A History of Walking* (New York: NYU Press, 2004), 6; and Robert Manning and Martha Manning, *Walks of a Lifetime* (Falcon Guides, 2017).

4   In a lifetime, the average: The average nondisabled American takes slightly more than 5,000 steps a day and has a life expectancy of seventy-nine years, meaning that most of us will take about 150 million steps. There are about 2,000 steps per mile, resulting in just under 75,000 miles. The circumference of the Earth is just under 25,000 miles, meaning that each of us, on average, will take enough steps to circle the Earth three times.

4   As primatologist John Napier wrote: John Napier, "The Antiquity of Human Walking," *Scientific American* 216, no. 4 (April 1967), 56–66.

5   By taking advantage of gravity: Timothy M. Griffin, Neil A. Tolani, and Rodger Kram, "Walking in Simulated Reduced Gravity: Mechanical Energy Fluctuations and Exchange," *Journal of Applied Physiology* 86, no. 1 (1999), 383–390.

6   In 2009, Jamaican sprinter: Dan Quarrell, "How Fast Does Usain Bolt Run in MPH/KM per Hour? Is He the Fastest Recorded Human Ever? 100m Record?" Eurosport.com, https://www.eurosport.com/athletics/how-fast-does-usain-bolt-run-in-mph-km-per-hour-is-he-the-fastest-recorded-human-ever-100m-record_sto5988142/story.shtml.

6   exceed sixty miles per hour: Cheetahs are often said to run seventy miles per hour, but the fastest ever recorded cheetah ran sixty-four miles per hour. N. C. C. Sharp, "Timed Running Speed of a Cheetah (*Acinonyx jubatus*)," *Journal of Zoology* 241, no. 3 (1997), 493–494.

6   According to the U.S. Centers: "Accidents or Unintentional Injuries," Centers for Disease Control and Prevention, National Center for Health Statistics, January 20, 2017, https://www.cdc.gov/nchs/fastats/accidental-injury.htm.

8   descended from apes: Humans are apes. We are a member of a family of large-bodied, fruit-eating, tailless primates called hominoids, which includes gorillas, chimpanzees, bonobos, orangutans, and gibbons. Hominoid is sometimes shorthanded to "ape." However, it is useful to have a word for us (human) and a word for nonhuman hominoids (ape). Even though I acknowl-

edge that we are, in fact, apes, throughout this book, I use the word "ape" as a substitute for nonhuman hominoid, and when I use it, I am referring to chimpanzees, gorillas, bonobos, orangutans, and/or gibbons.

8 "thrown on the origin of man": Throughout the book, I use the word "man" when it is a direct quote, as it is in this sentence from Darwin's *Origin of Species*, or when I am referring to actual men. This is not a useful or inclusive word to describe all humankind. The anthropologist Sally Linton (Slocum) wrote, "A theory that leaves out half of the human species is unbalanced" (in "Woman the Gatherer: Male Bias in Anthropology," in *Toward an Anthropology of Women*, ed. Rayna R. Reiter [New York: Monthly Review Press, 1975]). A word that does the same is similarly problematic.

8 Darwin predicted a century and a half ago: Charles Darwin wrote on p. 199 of *The Descent of Man*, "It is somewhat more probable that our early progenitors live on the African continent than elsewhere." He then wrote, "But it is useless to speculate on the subject."

8 The only premodern human fossils: In 1864, William King, a professor of geology in Ireland, named a new extinct human species on the basis of a partial skeleton from Feldhofer Cave in the Neander Valley in Germany. He called it *Homo neanderthalensis*. Neandertal fossils had also been found in Belgium and the Gibraltar peninsula. In 1864, Darwin even held the Gibraltar Neandertal in his hands but did not appreciate its significance. The Cro-Magnon *Homo sapiens* fossils were also known, having been discovered in 1868.

9 young Australian professor named Raymond Dart: See Raymond Dart, *Adventures with the Missing Link* (New York: Harper & Brothers, 1959), and Lydia Pyne, *Seven Skeletons* (New York: Viking, 2016), for more details on the backstory of Dart's discovery. In short, Dart's only female student, Josephine Salmons, spotted a baboon skull in the possession of a family friend, Mr. E. G. Izod. Izod was the director of the Northern Lime Company, which had been mining in the Buxton Limeworks Quarry in Taung, South Africa. Retellings of the story differ as to whether the fossil skull was on his mantelpiece or being used as a paperweight on his desk. Either way, Salmons brought the fossil to Dart. Dart was enthralled and reached out to Izod, requesting that additional fossils from the quarry be delivered to him for study. Dart recalls in his book that the boxes that contained the Taung child arrived the day he was in a tuxedo, hosting a friend's wedding.

9 he retrieved a small wooden box: In 1931, Dart brought the Taung child to London so that it could be studied by paleoanthropologists there. One day, Dart gave this box, with the Taung child inside, to his wife Dora to bring back to their apartment. But she mistakenly left it in a taxi. It spent much of the day riding around London, until the taxi driver noticed the box, opened it, and was shocked to find a child's skull inside! He immediately brought it to the police. Dora, by this time, had realized the box was missing and went to the London police, where she reclaimed the irreplaceable fossil. Close call.

10 2.5-million-year-old child's eye sockets: The geological age of Taung is uncertain. McKee (1993) dates it to 2.6–2.8 million years old. More recently, Kuhn et al. (2016) age Taung to 2.58–3.03 million years old. Jeffrey K. McKee, "Faunal Dating of the Taung Hominid Fossil Deposit," *Journal of Human Evolution* 25, no. 5 (1993), 363–376. Brian F. Kuhn et al., "Renewed Investigations

at Taung; 90 Years After the Discovery of *Australopithecus africanus*," *Palae-ontologica africana* 51 (2016), 10–26.

10 He called it *Australopithecus africanus*: Raymond A. Dart, "*Australopithecus africanus*: The Man-Ape of South Africa," *Nature* 115 (1925), 195–199.

11 these fossils are between 2.0 and 2.6 million years old: Robyn Pickering and Jan D. Kramers, "Reappraisal of the Stratigraphy and Determination of New U-Pb Dates for the Sterkfontein Hominin Site, South Africa," *Journal of Human Evolution* 59, no. 1 (2010), 70–86.

12 Makapansgat hominin *Australopithecus prometheus*: Raymond A. Dart, "The Makapansgat Proto-human *Australopithecus prometheus*," *American Journal of Physical Anthropology* 6, no. 3 (1948), 259–284.

12 In 1949, Dart published his findings: Raymond A. Dart, "The Predatory Implemental Technique of *Australopithecus*," *American Journal of Physical Anthropology* 7, no. 1 (1949), 1–38. The term "osteodontokeratic" appeared in 1957.

12 He had spent much of 1918 in England and France: Dart served as a medical officer at the Royal Prince Alfred Hospital before he was promoted to captain in the Australian Army Medical Corps (1918–1919). While I speculate that Dart may have seen the effects of the war, he never directly saw any action and wrote nothing that I could find about his experiences during World War I. See Phillip V. Tobias, "Dart, Raymond Arthur (1893–1988)," *Australian Dictionary of Biography*, vol. 17 (2007).

12 international bestseller *African Genesis*: Robert Ardrey, *African Genesis* (New York: Atheneum, 1961).

13 Dart's former student Phillip Tobias: Phillip Tobias would have a long and celebrated career, remaining active until his death in 2012. He excavated at Sterkfontein, worked with Louis Leakey in naming *Homo habilis*, and trained Lee Berger, who becomes an important part of this book in Chapters 7 and 9. Tobias fought against the apartheid regime from within South Africa, speaking at protest rallies for equal treatment of all South Africans. By the time I met him, the already-short Tobias had shrunk a few more inches and walked with a cane. He was wise and kind. I thought of him as paleoanthropology's Yoda.

13 *Australopithecus prometheus* and *africanus*: The scientific way to write a species name is to capitalize the genus, lowercase the species, and to write it in italics. Thus, we humans are *Homo sapiens*. The Taung child is *Australopithecus africanus*. To avoid writing *Australopithecus* over and over again, the proper way to abbreviate a species would be to write the first initial of the genus and then the species. Thus, we are *H. sapiens* and Taung is *A. africanus*. However, throughout this book, I have taken the liberty to shorten species even more and drop the genus, referring to them as *africanus*, *afarensis*, or *sapiens*. Scientifically, this is a no-no, but for readability, it makes more sense to bend the rules of taxonomic nomenclature.

13 *prometheus* was absorbed into *africanus*: John T. Robinson, "The Genera and Species of the Australopithecinae," *American Journal of Physical Anthropology* 12, no. 2 (1954), 181–200. On the basis of the partial skeleton StW 573, nicknamed "Little Foot," Ron Clarke has resurrected the species *Australopithecus prometheus*. This is controversial, however, and it is an open question whether the fossils from Sterkfontein and Makapansgat represent a single, variable species, or whether there are two different species of *Australopithecus*

in the sample. See Ronald J. Clarke, "Excavation, Reconstruction and Taphonomy of the StW 573 *Australopithecus prometheus* Skeleton from Sterkfontein Caves, South Africa," *Journal of Human Evolution* 127 (2019), 41–53. Ronald J. Clarke and Kathleen Kuman, "The Skull of StW 573, a 3.67 Ma *Australopithecus prometheus* Skeleton from Sterkfontein Caves, South Africa," *Journal of Human Evolution* 134 (2019), 102634.

13   catalogue name SK 54: Charles K. Brain, "New Finds at the Swartkrans Australopithecine Site," *Nature* 225 (1970), 1112–1119.

13   I traveled to the Ditsong museum: At the time of my visit, it was still called the Transvaal Museum. The Transvaal was the name of the South African province that included Pretoria (the administrative capital city) and Johannesburg from 1910 to 1994. With the fall of the apartheid regime, part of the district was renamed Gauteng, which means "place of gold" in the Sotho language. The museum was renamed "Ditsong," a Tswana word meaning "a place of heritage," in 2010.

13   The collections manager, Stephany Potze: Since 2016, Stephany Potze is no longer at the Ditsong National Museum of Natural History, but now is the lab manager for the La Brea Tar Pits and Museum in Los Angeles, California.

13   The Broom Room: SK 48 is a heavy, limestone-infused skull of a *Paranthropus robustus*, discovered at Swartkrans by Broom and J. T. Robinson in 1949. Sts 5, or Mrs. Ples, was found at Sterkfontein by Broom and Robinson in 1947 and is one of the best-preserved skulls of an adult *Australopithecus africanus*.

14   the lower jaw of an ancient leopard: The jaw has the catalogue number SK 349.

14   They were hunted: Charles K. Brain, *The Hunters or the Hunted? An Introduction to African Cave Taphonomy* (Chicago: University of Chicago Press, 1981). Also see Donna Hart and Robert W. Sussman, *Man the Hunted: Primates, Predators, and Human Evolution* (New York: Basic Books, 2005).

14   Some scholars doubt: The best example of this can be found in Matt Carmill, "Human Uniqueness and Theoretical Content in Paleoanthropology," *International Journal of Primatology* 11 (1990), 173–192.

CHAPTER 2: *T. REX*, THE CAROLINA BUTCHER, AND THE FIRST BIPEDS

17   "Four legs good": George Orwell, *Animal Farm* (London: Secker & Warburg, 1945).

17   120-million-year-old trail of footprints: Hang-Jae Lee, Yuong-Nam Lee, Anthony R. Fiorillo, and Junchang Lü, "Lizards Ran Bipedally 110 Million Years Ago," *Scientific Reports* 8, no. 2617 (2018), https://doi.org/10.1038/s41598-018-20809-z. The date of the tracks is between 110 million and 128 million years old.

18   *Eudibamus* is one of the earliest known: David S. Berman et al., "Early Permian Bipedal Reptile," *Science* 290, no. 5493 (2000), 969–972. *Cabarzia trostheidei* was discovered in Germany in 2019 and is 15 million years older than *Eudibamus*. Frederik Spindler, Ralf Werneburg, and Joerg W. Schneider, "A New Mesenosaurine from the Lower Permian of Germany and the Postcrania of *Mesenosaurus*: Implications for Early Amniote Comparative Ostology," *Paläontologische Zeitschrift* 93 (2019), 303–344.

19   our feathered friends are most closely related to crocodilians: See Axel Janke
      and Ulfur Arnason, "The Complete Mitochondrial Genome of *Alligator mis-
      sissippiensis* and the Separation Between Recent Archosauria (Birds and
      Crocodiles)," *Molecular Biology and Evolution* 14, no. 12 (1997), 1266–1272,
      and Richard E. Green et al., "Three Crocodilian Genomes Reveal Ances-
      tral Patterns of Evolution Among Archosaurs," *Science* 346, no. 6215 (2014),
      1254449. A colleague of mine pointed out that comparative anatomists and
      paleontologists have long known that birds and crocodiles are related and
      did not need genetics to tell us that. See Robert L. Carroll, *Vertebrate Paleon-
      tology and Evolution* (New York: W. H. Freeman, 1988).

19   "This is what evolutionists": "God Must Exist . . . Because the Crocoduck
      Doesn't," *Nightline Face-off with Martin Bashir*, ABC News, https://www.you
      tube.com/watch?v=a0DdgSDan9c. Funny thing, though, a Cretaceous croc-
      odile discovered in the early 2000s had a duck bill and probably skimmed
      the water for food like ducks do. It was named *Anatosuchus*, which means
      "crocoduck." Paul Sereno, Christian A. Sidor, Hans C. E. Larsson, and
      Boubé Gado, "A New Notosuchian from the Early Cretaceous of Niger,"
      Journal of Vertebrate Paleontology 23, no. 2 (2003), 477–482.

20   She called it *Carnufex carolinensis*: Lindsay E. Zanno, Susan Drymala, Ster-
      ling J. Nesbit, and Vincent P. Schneider, "Early Crocodylomorph Increases
      Top Tier Predator Diversity During Rise of Dinosaurs," *Scientific Reports* 5
      (2015), 9276. See also Susan M. Drymala and Lindsay E. Zanno, "Osteology
      of *Carnufex carolinensis* (Archosauria: Pseudosuchia) from the Pekin Forma-
      tion of North Carolina and Its Implications for Early Crocodylomorph Evo-
      lution," *PLOS ONE* 11, no. 6 (2016), e0157528.

20   the earliest crocodile ancestors were lightly built: In 2020, researchers de-
      scribed fossil footprints left by bipedal walking crocodiles in 106-million-
      year-old sediments in South Korea. See Kyung Soo Kim, Martin G. Lockley,
      Jong Deock Lim, Seul Mi Bae, and Anthony Romilio, "Trackway Evidence
      for Large Bipedal Crocodylomorphs from the Cretaceous of Korea," *Scientific
      Reports* 10, no. 8680 (2020).

22   From the hingelike shape: From Riley Black (formerly Brian Switek), *My Beloved
      Brontosaurus* (New York: Scientific American/Farrar, Straus & Giroux, 2013).

24   whether they served any function: In his book, Steve Brusatte discusses the
      work of colleague Sara Burch, who determined that *T. rex* arms were "accesso-
      ries to murder." Like giant meat hooks, they would have held on to prey trying
      to escape the jaws of a *T. rex*. Steve Brusatte, *The Rise and Fall of the Dinosaurs:
      The Untold Story of a Lost World* (New York: William Morrow, 2018), 215.

26   Researchers from the University of Alberta: W. Scott Persons and Philip J.
      Currie, "The Functional Origin of Dinosaur Bipedalism: Cumulative Evi-
      dence from Bipedally Inclined Reptiles and Disinclined Mammals," *Jour-
      nal of Theoretical Biology* 420, no. 7 (2017), 1–7. Persons wrote to me in an
      email, "The big tail muscles aren't unique to bipedal dinosaurs (nearly all di-
      nosaurs have them). But having the tail muscles means that, when you do
      start to evolve towards speed, you are naturally inclined to go bipedal." In
      other words, because of these muscles, the back legs outperform the front. To
      maximize the power of the tail muscles, selection would therefore favor elon-
      gated back legs in fast dinosaurs and front legs that are tucked out of the way.

26 Picture the posture: Turns out, *T. rex* probably could not have run as quickly as Hollywood would have us believe. See Brusatte, *The Rise and Fall of the Dinosaurs*, 210–212.

27 Monkeys typically cannot do this: Exceptions to this are the South American atelid monkeys, which through convergent evolution have obtained apelike shoulder mobility. These include spider monkeys, howler monkeys, woolly monkeys, and muriquis.

27 closest large landmass—Australia: This landmass, which connected mainland Australia to Tasmania and New Guinea, is called Sahul.

28 this is a very effective form of locomotion: Robert McN. Alexander and Alexandra Vernon, "The Mechanics of Hopping by Kangaroos (Macropodidae)," *Journal of Zoology* 177, no. 2 (1975), 265–303.

29 skull of the Luck Dragon: As I would discover later, Riley Black joked in a blog on the ten best fossil mammals that *Andrewsarchus* was a "real life version of Gmork from *The Neverending Story*." See https://www.tor.com/2015/01/04/ten-fossil-mammals-as-awesome-as-any-dinosaur-2. In an email, Black called this a case of convergent comedic evolution!

29 Christine Janis, a paleontologist: Christine M. Janis, Karalyn Buttrill, and Borja Figueirido, "Locomotion in Extinct Giant Kangaroos: Were Sthenurines Hop-Less Monsters?" *PLOS ONE* 9, no. 10 (2014), e109888.

29 Four-million-year-old footprints: Aaron B. Camens and Trevor H. Worthy, "Walk Like a Kangaroo: New Fossil Trackways Reveal a Bipedally Striding Macropodid in the Pliocene of Central Australia," *Journal of Vertebrate Paleontology* (2019), 72.

30 While mostly quadrupedal, there is footprint: Footprints found at the site of Pehuén-Có, Argentina, have indicated to some researchers a slow, bipedal gait for *Megatherium*. R. Ernesto Blanco and Ada Czerwonogora, "The Gait of *Megatherium* CUVIER 1796 (Mammalia, Xenartha, Megatheriidae)," *Senckenbergiana Biologica* 83, no. 1 (2003), 61–68. Another team attributes the bipedal prints to *Neomegatherichnum pehuencoensis*, a different type of giant sloth. Silvia A. Aramayo, Teresa Manera de Bianco, Nerea V. Bastianelli, and Ricardo N. Melchor, "Pehuen Co: Updated Taxonomic Review of a Late Pleistocene Ichnological Site in Argentina," *Palaeogeography, Palaeoclimatology, Palaeoecology* 439 (2015), 144–165.

30 no larger than chimpanzees: Mark Grabowski and William L. Jungers, "Evidence of a Chimpanzee-Sized Ancestor of Humans but a Gibbon-Sized Ancestor of Apes," *Nature Communications* 8, no. 880 (2017).

CHAPTER 3: "HOW THE HUMAN STOOD UPRIGHT" AND OTHER JUST-SO STORIES ABOUT BIPEDALISM

31 "Speculations on the origins": Jonathan Kingdon, *Lowly Origin: When, Where, and Why Our Ancestors First Stood Up* (Princeton, NJ: Princeton University Press, 2003), 16.

31 This worried Zeus: Plato, *The Symposium*, trans. Christopher Gill (New York: Penguin Classics, 2003).

32 University of Chicago anthropologist: Russell H. Tuttle, David M. Webb, and Nicole I. Tuttle, "Laetoli Footprint Trails and the Evolution of Hominid

Bipedalism," in *Origine(s) de la Bipédie chez les Hominidés*, ed. Yves Coppens and Brigitte Senut (Paris: Éditions du CNRS, 1991), 187–198.

32   The question, then: Napier (1964) wrote: "Occasional bipedalism is almost the rule among Primates." John R. Napier, "The Evolution of Bipedal Walking in the Hominids," *Archives de Biologie (Liège)* 75 (1964), 673–708. In other words, the capacity is there to a degree, but the incentive often is not. Paleontologist Mike Rose also argued that bipedalism was part of the locomotor repertoire of the last common ancestor and what is at issue is what caused the increase in frequency of the behavior in hominins. Michael D. Rose, "The Process of Bipedalization in Hominids," in *Origine(s) de la Bipédie chez les Hominidés*, eds. Yves Coppens and Brigitte Senut (Paris: Éditions du CNRS, 1991), 37–48. Anthropologist Jon Marks also pointed out that this is not something new, but the evolution of exclusive bipedalism. He would argue that behavior preceded morphology, making bipedalism Lamarckian to some degree. Jon Marks, "Genetic Assimilation in the Evolution of Bipedalism," *Human Evolution* 4, no. 6 (1989), 493–499. Tuttle also argued that "bipedalism preceded the emergence of the Hominidae," given that every ape is occasionally bipedal. Russell H. Tuttle, "Evolution of Hominid Bipedalism and Prehensile Capabilities," *Philosophical Transactions of the Royal Society of London B* 292 (1981), 89–94.

33   These examples support the "peekaboo": Tuttle came up with other great names for the various hypotheses, including schlepp, trenchcoat, all wet, tagalong, hot to trot, two feet are better than four, swingers go further, upward mobility, and hit 'em where it hurts. Tuttle, Webb, and Tuttle, "Laetoli Footprint Trails," 187–198.

33   He wrote that bipedalism satisfied: Jean-Baptiste Lamarck, *Zoological Philosophy, or Exposition with Regard to the Natural History of Animals* (Paris: Musée d'Histoire Naturelle, 1809).

33   Perhaps our ancient hominin: Nina G. Jablonski and George Chaplin, "Origin of Habitual Terrestrial Bipedalism in the Ancestor of the Hominidae," *Journal of Human Evolution* 24, no. 4 (1993), 259–280.

33   One scholar took this a step further: A. Kortlandt, "How Might Early Hominids Have Defended Themselves Against Large Predators and Food Competitors?" *Journal of Human Evolution* 9 (1980), 79–112.

34   In the wild, some chimpanzees: Kevin D. Hunt, "The Evolution of Human Bipedality: Ecology and Functional Morphology," *Journal of Human Evolution* 26, no. 3 (1994), 183–202. Craig B. Stanford, *Upright: The Evolutionary Key to Becoming Human* (New York: Houghton Mifflin Harcourt, 2003). Craig B. Stanford, "Arboreal Bipedalism in Wild Chimpanzees: Implications for the Evolution of Hominid Posture and Locomotion," *American Journal of Physical Anthropology* 129, no. 2 (2006), 225–231.

34   Still others position our ancestors: Richard Wrangham, Dorothy Cheney, Robert Seyfarth, and Esteban Sarmiento, "Shallow-Water Habitats as Sources of Fallback Foods for Hominins," *American Journal of Physical Anthropology* 140, no. 4 (2009), 630–642.

34   This hypothesis is a more reasonable: Sir Alister Hardy, "Was Man More Aquatic in the Past?" *New Scientist* (March 17, 1960). Elaine Morgan, *The Aquatic Ape: A Theory of Human Evolution* (New York: Stein & Day, 1982).

Elaine Morgan, *The Aquatic Ape Hypothesis: Most Credible Theory of Human Evolution* (London: Souvenir Press, 1999). Morgan's TED Talk, "I Believe We Evolved from Aquatic Apes," TED.com, https://www.ted.com/talks/elaine _morgan_i_believe_we_evolved_from_aquatic_apes. David Attenborough, "The Waterside Ape," BBC Radio, https://www.bbc.co.uk/programmes /b07v0hhm. See also Marc Verhaegen, Pierre-François Puech, and Stephen Murro, "Aquarboreal Ancestors?" *Trends in Ecology & Evolution* 17, no. 5 (2002), 212–217. Algis Kuliukas, "Wading for Food the Driving Force of the Evolution of Bipedalism?" *Nutrition and Health* 16 (2002), 267–289.

35   The aquatic-ape hypothesis: What this hypothesis lacks in data it makes up for in a relentless marketing campaign. Proponents of this idea use Twitter, email, the comments section of YouTube, and Amazon book reviews to promote some version of the aquatic-ape hypothesis. For example, I encountered dozens of Amazon book reviews giving only two stars to books, including textbooks, that fail to adopt the aquatic-ape hypothesis as *the* explanation for bipedal origins. For a dismantling of the aquatic ape, see John H. Langdon, "Umbrella Hypotheses and Parsimony in Human Evolution: A Critique of the Aquatic Ape Hypothesis," *Journal of Human Evolution* 33, no. 4 (1997), 479–494. For information on human water use, see Herman Pontzer et al., "Evolution of Water Conservation in Humans," *Current Biology* 31, no. 8 (2021), 1804-1810.e5

36   But everyone loves a good mystery: Björn Merker, "A Note on Hunting and Hominid Origins," *American Anthropologist* 86, no. 1 (1984), 112–114. Kingdon, *Lowly Origin* (2003). R. D. Guthrie, "Evolution of Human Threat Display Organs," *Evolutionary Biology* 4, no. 1 (1970), 257–302. David R. Carrier, "The Advantage of Standing Up to Fight and the Evolution of Habitual Bipedalism in Hominins," *PLOS ONE* 6, no. 5 (2011), e19630. Uner Tan, "Two Families with Quadrupedalism, Mental Retardation, No Speech, and Infantile Hypotonia (Uner Tan Syndrome Type-II): A Novel Theory for the Evolutionary Emergence of Human Bipedalism," *Frontiers in Neuroscience* 8, no. 84 (2014), 1–14. Anthony R. E. Sinclair, Mary D. Leakey, and M. Norton-Griffiths, "Migration and Hominid Bipedalism," *Nature* 324 (1986), 307–308. Edward Reynolds, "The Evolution of the Human Pelvis in Relation to the Mechanics of the Erect Posture," *Papers of the Peabody Museum of American Archaeology and Ethnology* 11 (1931), 255–334. Isabelle C. Winder et al., "Complex Topography and Human Evolution: The Missing Link," *Antiquity* 87, no. 336 (2013), 333–349. Milford H. Wolpoff, *Paleoanthropology* (New York: McGraw-Hill College, 1998). Sue T. Parker, "A Sexual Selection Model for Hominid Evolution," *Human Evolution* 2 (1987), 235–253. Adrian L. Melott and Brian C. Thomas, "From Cosmic Explosions to Terrestrial Fires," *Journal of Geology* 127, no. 4 (2019), 475–481.

37   Ostrich Mimicry: See also Carolyn Brown, "IgNobel (2): Is That Ostrich Ogling Me?" *Canadian Medical Association Journal* 167, no. 12 (2002), 1348.

37   And there are many, many more: And more reasons to be skeptical of them. In 2008, Ken Sayers and C. Owen Lovejoy adopted a philosophy known as "Jolly's paradox" to argue against using bipedal behavior in other primates to speculate on its origins in hominins. They argued that the circumstances behind bipedal locomotion in other primates cannot be the reasons hominins

began moving on two legs, otherwise these other primates would also have adopted full-time upright walking as a way to move. Ken Sayers and C. Owen Lovejoy, "The Chimpanzee Has No Clothes: A Critical Examination of *Pan troglodytes* in Models of Human Evolution," *Current Anthropology* 49, no. 1 (2008), 87–114.

38 Molecular anthropologist Todd Disotell: Not long after our interview, Disotell accepted a new position at the University of Massachusetts, Amherst.

40 our lineage had fully split: I insert the word "fully" here because lineages rarely experience rapid speciation, and instead the splitting of lineages is often a slow, messy process in which hybridization continues to occur before lineages become reproductively isolated. See Nick Patterson, Daniel J. Richter, Sante Gnerre, Eric S. Lander, and David Reich, "Genetic Evidence for Complex Speciation of Humans and Chimpanzees," *Nature* 441 (2006), 1103–1108. Alywyn Scally et al., "Insights into Hominid Evolution from the Gorilla Genome Sequence," *Nature* 483 (2012), 169–175. Furthermore, there is a downstream effect of a deeper (i.e., 12-million-year) human-chimpanzee divergence, which would push the monkey-ape divergence into the early Oligocene. This is at odds with the fossil record, which has produced evidence of common ape-monkey ancestors as early as 29 million years ago.

42 However, we are not sure *why* upright: Anthropologists Henry McHenry and Peter Rodman said that bipedalism was "an ape's way of living where an ape could not live." Roger Lewin, "Four Legs Bad, Two Legs Good," *Science* 235 (1987), 969–971.

42 One explanation is that upright posture: Peter E. Wheeler, "The Evolution of Bipedality and the Loss of Functional Body Hair in Hominids," *Journal of Human Evolution* 13, no. 1 (1984), 91–98. Peter E. Wheeler, "The Thermoregulatory Advantages of Hominid Bipedalism in Open Equatorial Environments: The Contribution of Increased Convective Heat Loss and Cutaneous Evaporative Cooling," *Journal of Human Evolution* 21, no. 2 (1991), 107–115.

42 Researchers from Harvard University: Michael D. Sockol, David A. Raichlen, and Herman Pontzer, "Chimpanzee Locomotor Energetics and the Origin of Human Bipedalism," *Proceedings of the National Academy of Sciences* 104, no. 30 (2007), 12265–12269.

43 twice as much energy: In the original Sockol et al. 2007 paper, the researchers reported a value of four times as much energy use in chimpanzees as in humans. This number has since been updated to twice as much energy. See Herman Pontzer, David A. Raichlen, and Michael D. Sockol, "The Metabolic Cost of Walking in Humans, Chimpanzees, and Early Hominins," *Journal of Human Evolution* 56, no. 1 (2009), 43–54. Herman Pontzer, David A. Raichlen, and Peter S. Rodman, "Bipedal and Quadrupedal Locomotion in Chimpanzees," *Journal of Human Evolution* 66 (2014), 64–82.

43 there was nothing energetically special: See Herman Pontzer, "Economy and Endurance in Human Evolution," *Current Biology* 27, no. 12 (2017), R613–R621. Lewis Halsey and Craig White, "Comparative Energetics of Mammalian Locomotion: Humans Are Not Different," *Journal of Human Evolution* 63 (2012), 718–722.

44 Susana Carvalho, an anthropologist: Susana Carvalho et al., "Chimpanzee Carrying Behaviour and the Origins of Human Bipedality," *Current Biology*

22, no. 6 (2012), R180–R181. There are two kinds of nuts that the chimpanzees consume and that I combine under the name "African walnuts"—oil palm nuts (*Elaeis guineensis*) and coula nuts (*Coula edulis*).

44   This idea goes back to Gordon Hewes: Gordon W. Hewes, "Food Transport and the Origin of Hominid Bipedalism," *American Anthropology* 63, no. 4 (1961), 687–710. Gordon W. Hewes, "Hominid Bipedalism: Independent Evidence for the Food-Carrying Theory," *Science* 146, no. 3642 (1964), 416–418.

45   This "provisioning hypothesis": C. Owen Lovejoy, "The Origin of Man," *Science* 211, no. 4480 (1981), 341–350. C. Owen Lovejoy, "Reexamining Human Origins in Light of *Ardipithecus ramidus*," *Science* 326, no. 5949 (2009), 74–74e8.

45   To many critics of this idea: See papers in Lori Hager, *Women in Human Evolution* (New York: Routledge, 1997).

45   In the 1970s and 1980s, anthropologists Nancy: Nancy Tanner and Adrienne Zihlman, "Women in Evolution, Part I: Innovation and Selection in Human Origins," *Signs* 1, no. 3 (1976), 585–605. Adrienne Zihlman, "Women in Evolution, Part II: Subsistence and Social Organization Among Early Hominids," *Signs* 4, no. 1 (1978), 4–20. Nancy M. Tanner, *On Becoming Human* (Cambridge: Cambridge University Press, 1981).

46   It has been shown that in bonobos: Thibaud Gruber, Zanna Clay, and Klaus Zuberbühler, "A Comparison of Bonobo and Chimpanzee Tool Use: Evidence for a Female Bias in the *Pan* Lineage," *Animal Behavior* 80, no. 6 (2010), 1023–1033. Many of the innovating primates discussed are female in Frans de Waal, *The Ape and the Sushi Master: Cultural Reflections of a Primatologist* (New York: Basic Books, 2008). Klaree J. Boose, Frances J. White, and Audra Meinelt, "Sex Differences in Tool Use Acquisition in Bonobos (*Pan paniscus*)," *American Journal of Primatology* 75, no. 9 (2013), 917–926. At Fongoli, a study site in Senegal,chimpanzees—mostly females—hunt with sharpened sticks. Jill D. Pruetz et al., "New Evidence on the Tool-Assisted Hunting Exhibited by Chimpanzees (*Pan troglodytes verus*) in a Savannah Habitat at Fongoli, Sénégal," *Royal Society of Open Science* 2 (2015), 140507.

CHAPTER 4: LUCY'S ANCESTORS

47   "But we must not fall": Charles Darwin, *The Descent of Man, and Selection in Relation to Sex*, vol. I (London: John Murray, 1871), 199.

47   The leg indicated: It appears that Dubois was right, but for the wrong reasons. In 2015, Chris Ruff and colleagues reexamined the Trinil femur and concluded that it derived from a much more recent time period than the skull, and likely belongs to *Homo sapiens*. However, Dubois discovered four other femurs at Trinil in 1900 and described them in the 1930s. Ruff's reassessment of them found that they are consistent with the anatomy of *Homo erectus*. Thus, Dubois's conclusion that *Pithecanthropus erectus* was bipedal turns out to have been based on a *Homo sapiens* femur, though additional femurs he discovered show he was right. Christopher B. Ruff, Laurent Puymerail, Roberto Machiarelli, Justin Sipla, and Russell L. Ciochon, "Structure and Composition of the Trinil Femora: Functional and Taxonomic Implications," *Journal of Human Evolution* 80 (2015), 147–158.

48   In 1900, Dubois and his son: See Pat Shipman, *The Man Who Found the Missing Link: Eugène Dubois and His Lifelong Quest to Prove Darwin Right* (Cambridge, MA: Harvard University Press, 2002).

48   But the way Boule interpreted the body: The La Chapelle individual was elderly and arthritic when he died. In life, therefore, he was hunched over not because the species lacked a fully erect posture, but rather because La Chapelle had lived long enough to develop a pathological skeleton.

49   Arizona State University paleoanthropologist: At the time of the discovery, Donald Johanson's appointment was with the Cleveland Museum of Natural History. Throughout this book, I generally try to acknowledge where scientists are currently working rather than where they were when the work being discussed was done.

49   and his colleagues: Throughout this book, I do my best to acknowledge the great work being done by my fellow scientists. However, science is rarely done alone, and usually large teams contribute to all of the studies I discuss. In these pages of endnotes, the term "et al.," which is used for any study with more than five authors, appears over 120 times. I'm inspired to mention this by Robert Sapolsky, who, in the footnotes of his recent book *Behave*, wrote, "Whenever I describe work done by Jane Doe or Joe Smith, I actually mean 'work done by Doe and a team of her postdocs, technicians, grad students, and collaborators spread far and wide over the years.' I'll be referring solely to Doe or Smith for brevity, not to imply that they did all the work on their own—science is utterly a team process."

49   Upon the umpteenth playing: Donald C. Johanson, *Lucy: The Beginnings of Humankind* (New York: Simon & Schuster, 1981).

50   It is unclear how she died: See John Kappelman et al., "Perimortem Fractures in Lucy Suggest Mortality from Fall out of Tall Tree," *Nature* 537 (2016), 503–507.

52   human babies also have a spine shaped this way: This, it turns out, is complicated. Human infants are already born with some S-shaped curvature of the spine. See Elie Choufani et al., "Lumbosacral Lordosis in Fetal Spine: Genetic or Mechanic Parameter," *European Spine Journal* 18 (2009), 1342–1348. However, the spine becomes more lordotic developmentally, particularly at the age when kids begin to take their first steps. M. Maurice Abitbol, "Evolution of the Lumbosacral Angle," *American Journal of Physical Anthropology* 72, no. 3 (1987), 361–372. But it appears as though this would happen no matter what. Children who never walk still develop an S-shaped spine. Sven Reichmann and Thord Lewin, "The Development of the Lumbar Lordosis," *Archiv für Orthopädische und Unfall-Chirurgie, mit Besonderer Berücksichtigung der Frakturenlehre und der Orthopädisch-Chirurgischen Technik* 69 (1971), 275–285.

52   muscles called the lesser gluteals: The muscles I'm referring to here are gluteus medius and gluteus minimus, the so-called lesser gluteals, compared with the much more massive gluteus maximus, our butt muscle.

53   she walked around her world on two legs: See C. Owen Lovejoy, "Evolution of Human Walking," *Scientific American* (November 1988), 118–125. When I write that the hips are on the side of the body, this is shorthand for the iliac blades having rotated to the side of the body, where they are in humans,

in contrast to apes who possess flat iliac blades that face the back of the body.

53   Chimpanzees never develop one: Christine Tardieu, "Ontogeny and Phylogeny of Femoro-Tibial Characters in Humans and Hominid Fossils: Functional Influence and Genetic Determinism," *American Journal of Physical Anthropology* 110 (1999), 365–377.

53   However, the year before Lucy: This specimen has the catalogue number A.L. 129–1 and is a different individual from Lucy. Details of discovery and the importance of the anatomy can be found in Johanson, *Lucy: The Beginnings of Humankind*. Donald C. Johanson and Maurice Taieb, "Plio-Pleistocene Hominid Discoveries in Hadar, Ethiopia," *Nature* 260 (1976), 293–297.

54   her head was never found: Probably. Francis Thackeray has raised the possibility that the head of the partial skeleton Sts 14 is Sts 5, Mrs. Ples. He also has proposed that Mrs. Ples is a juvenile male. Francis Thackeray, Dominique Gommery, and Jose Braga, "Australopithecine Postcrania (Sts 14) from the Sterkfontein Caves, South Africa: The Skeleton of 'Mrs Ples'?" *South African Journal of Science* 98, no. 5–6 (2002), 211–212. But also see Alejandro Bonmatí, Juan-Luis Arsuaga, and Carlos Lorenzo, "Revisiting the Developmental Stage and Age-at-Death of the 'Mrs. Ples' (Sts 5) and Sts 14 Specimens from Sterkfontein (South Africa): Do They Belong to the Same Individual?" *Anatomical Record* 291, no. 12 (2008), 1707–1722.

55   Often, a trade-off: Credit for planting this idea in my brain goes to Boston University geologist Andy Kurtz, with whom I had the pleasure of coteaching in the winter of 2015.

56   ^{40}K and ^{40}Ar: Throughout this section, I write about potassium and argon. However, researchers have developed a shortcut that improves the accuracy of this technique, known as ^{40}Ar/^{39}Ar (argon-argon) dating.

56   Because her bones: Robert C. Walter, "Age of Lucy and the First Family: Single-Crystal ^{40}Ar/^{39}Ar Dating of the Denen Dora and Lower Kada Hadar Members of the Hadar Formation, Ethiopia," *Geology* 22, no. 1 (1994), 6–10.

57   At a state dinner: Juliet Eilperin, "In Ethiopia, Both Obama and Ancient Fossils Get a Motorcade," *Washington Post*, July 27, 2015.

58   In the mid-1990s, however, Meave: Meave G. Leakey, Craig S. Feibel, Ian McDougall, and Alan Walker, "New Four-Million-Year-Old Hominid Species from Kanapoi and Allia Bay, Kenya," *Nature* 376 (1995), 565–571.

60   The anatomy of these hominin fossils: Brigitte Senut et al., "First Hominid from the Miocene (Lukeino Formation, Kenya)," *Comptes Rendus de l'Académie des Sciences—Series IIA—Earth and Planetary Science* 332, no. 2 (2001), 137–144.

60   If Senut and Pickford had found: I studied casts of *Orrorin tugenensis* in Senut and Pickford's lab in the fall of 2019. What is preserved of the most complete femur has all of the hallmarks of an upright walking hominin. From this bone alone, I, too, would have concluded—as these researchers did—that *Orrorin* was bipedal. I'm eager to see what the rest of this hominin looked like!

60   The details, which include claims of forged: See Ann Gibbons, *The First Human: The Race to Discover Our Earliest Ancestors* (New York: Anchor Books, 2007). Perhaps the quote that best sums up the saga of the *Orrorin* fossils was uttered by Brigitte Senut, who had an unexpected response to the discovery

of the oldest hominin femur: "I told Martin to throw it in the lake. It would only bring us trouble." (From Gibbons, p. 195.)

60   They are rumored: In 2018, I corresponded with Eustace Gitonga, the director of the Community Museums of Kenya (CMK), who is in possession of the *Orrorin* fossils. I requested to study the *Orrorin* material and was told that "the original *Orrorin* fossils are unavailable until the details of the new MOU are finalized." Here, Gitonga is referring to a memorandum of understanding between the CMK and the Baringo County government, which, according to Gitonga, feel as though foreign researchers have reneged on previous MOUs.

61   Only six months after *Orrorin*: Yohannes Haile-Selassie, "Late Miocene Hominids from the Middle Awash, Ethiopia," *Nature* 412 (2001), 178–181.

63   It had a combination of anatomies: Michel Brunet et al., "A New Hominid from the Upper Miocene of Chad, Central Africa," *Nature* 418 (2002), 145–151. Patrick Vignaud et al., "Geology and Palaeontology of the Upper Miocene Toro-Menalla Hominid Locality, Chad," *Nature* 418 (2002), 152–155.

64   After all, the skull was quite crushed: Milford Wolpoff, Brigitte Senut, Martin Pickford, and John Hawks, "Palaeoanthropology (Communication Arising): *Sahelanthropus* or '*Sahelpithecus*'?" *Nature* 419 (2002), 581–582. Brunet et al., "Reply," *Nature* 419 (2002), 582. Milford Wolpoff, John Hawks, Brigitte Senut, Martin Pickford, and James Ahern, "An Ape or *the* Ape: Is the Toumaï Cranium TM 266 a Hominid?" *PaleoAnthropology* (2006), 35–50.

64   The result appeared to show a humanlike hole: Christoph P. E. Zollikofer et al., "Virtual Cranial Reconstruction of *Sahelanthropus tchadensis*," *Nature* 434 (2005), 755–759. Franck Guy et al., "Morphological Affinities of the *Sahelanthropus tchadensis* (Late Miocene Hominid from Chad) Cranium," *Proceedings of the National Academy of Sciences* 105, no. 52 (2005), 18836–18841.

64   He found a femur: Though at the time it was not identified as a primate femur. In 2004, Aude Bergeret, then a graduate student at the University of Poitiers, was studying the faunal fossils from the Toros-Menalla locality when she identified the bone as belonging to a large primate. The only large primate known from Toros-Menalla is *Sahelanthropus tchadensis*. In 2018, Bergeret and her former mentor Roberto Macchiarelli proposed to present their work on the femur to the Anthropological Society of Paris, but, to the bewilderment of the entire paleoanthropological community, the abstract was rejected by the meeting organizers. See Ewen Callaway, "Controversial Femur Could Belong to Ancient Human Relative," *Nature* 553 (2018), 391–392.

64   In 2020, competing teams published preliminary analyses: Franck Guy et al., "Postcranial Evidence of Late Miocene Hominin Bipedalism in Chad," *Nature Research* (2020). Roberto Macchiarelli, Aude Bergeret-Medina, Damiano Marchi, and Bernard Wood, "Nature and Relationships of *Sahelanthropus tchadensis*," *Journal of Human Evolution* 149 (2020), 102898.

66   "No apology need be given": Robert Broom, "Further Evidence on the Structure of the South African Pleistocene Anthropoids," *Nature* 142 (1938), 897–899. More than a decade later, Broom and his student J. T. Robinson wrote: "In South Africa we have been making important discoveries so fast recently that it is quite impossible to publish memoirs on them within a year or even

two. We might withhold publication for many years as is so often done in the Northern Hemisphere, or we might issue preliminary descriptions and render ourselves liable to the criticism that our descriptions are inadequate. We think it much preferable to issue even inadequate descriptions and let other workers know something of our finds than to keep them secret for 10 years or more." Robert Broom and John T. Robinson, "Brief Communications: Notes on the Pelves of the Fossil Ape-Men," *American Journal of Physical Anthropology* 8, no. 4 (1950), 489–494. Four months later, Broom died at the age of eighty-four.

66   I look forward to the day: An educational supply company called Bone Clones, Inc., has sculpted versions of *Ardipithecus* and *Sahelanthropus* from published measurements and photographs. With these as our only options for hands-on teaching, many of us anthropologists have purchased the Bone Clones versions—for $295/skull—for our teaching labs. I, and my colleagues, would prefer to send that money to Chad and Ethiopia for actual casts of fossils, but that is not an option currently. Now that I have seen the original foot bones from *Ardipithecus ramidus* and proper casts of *Sahelanthropus*, I can report that the Bone Clones versions—despite their best efforts—are woefully inaccurate and, in some ways, even misleading. The actual *Sahelanthropus* cranium, for instance, is about 20 percent larger than the Bone Clones replica.

66   As one researcher put it: Daniel E. Lieberman, *The Story of the Human Body: Evolution, Health, and Disease* (New York: Pantheon, 2013), 33. He wrote, "You could fit all the fossils from *Ardipithecus*, *Sahelanthropus*, and *Orrorin* in a single shopping bag." I added the "and still have plenty of room for the groceries" part.

CHAPTER 5: ARDI AND THE RIVER GODS

67   "Let's just say *ramidus*": Rick Gore, "The First Steps," *National Geographic* (February 1997), 72–99.

67   In September 1994: Tim D. White, Gen Suwa, and Berhane Asfaw, "*Australopithecus ramidus*, a New Species of Early Hominid from Aramis, Ethiopia," *Nature* 371 (1994), 306–312. The type specimen—a collection of associated teeth—was found by a local Afar man, Gada Hamed.

67   But six months later: Tim D. White, Gen Suwa, and Berhane Asfaw, "Corrigendum: *Australopithecus ramidus*, a New Species of Early Hominid from Aramis, Ethiopia," *Nature* 375 (1995), 88.

68   Some referred to it as the Manhattan Project: Rex Dalton, "Oldest Hominid Skeleton Revealed," *Nature* (October 1, 2009). Donald Johanson and Kate Wong, *Lucy's Legacy: The Quest for Human Origins* (New York: Broadway Books, 2010), 154.

69   Ardi lived and died in a wooded environment: Giday WoldeGabriel et al., "The Geological, Isotopic, Botanical, Invertebrate, and Lower Vertebrate Surroundings of *Ardipithecus ramidus*," *Science* 326, no. 5949 (2009), 65–65e5. As with many seemingly simple statements in paleoanthropology, this one is contentious. Some scholars have argued that the Aramis locality would not have been as wooded as White and his colleagues suggest. Thure E. Cer-

ling et al., "Comment on the Paleoenvironment of *Ardipithecus ramidus,*" *Science* 328 (2010), 1105. Additionally, a second *Ardipithecus ramidus* fossil locality, at Gona, Ethiopia, does appear to be more of a grassland environment, implying that *Ardipithecus* was able to live in varied environments. Sileshi Semaw et al., "Early Pliocene Hominids from Gona, Ethiopia," *Nature* 433 (2005), 301–305. Interestingly, there is a second *Ardipithecus* partial skeleton from the more open Gona locality and it appears to have better-developed (i.e., more humanlike) skeletal adaptations for bipedal walking than the *Ardipithecus* from the more wooded Aramis locality. Scott W. Simpson, Naomi E. Levin, Jay Quade, Michael J. Rogers, and Sileshi Semaw, "*Ardipithecus ramidus* Postcrania from the Gona Project Area, Afar Regional State, Ethiopia," *Journal of Human Evolution* 129 (2019), 1–45.

70   They are delicate and not as dense: This refers to the fossils themselves. In life, Lucy and Ardi would have had similar bone density to their skeletons.

70   codirector with White: The late J. Desmond Clark formed this group in 1981. Other project directors are the previously mentioned geologist Giday WoldeGabriel and Yonas Beyene, who specializes in archaeology.

71   It was a compromise: In 2019, Scott Simpson of Case Western Reserve University published his team's analysis of another partial skeleton of an *Ardipithecus ramidus* discovered in the Gona region of Ethiopia. Although I have not yet studied the original fossil, it looks to me like it is even better adapted for bipedal locomotion than Ardi. If this turns out to be the case, then at this time (4.4 million years ago), there was variation in bipedal abilities in *Ardipithecus*—natural selection could have favored those individuals better suited for bipedal walking to eventually evolve a habitually bipedal *Australopithecus* from a facultatively bipedal *Ardipithecus.*

72   This image, called the March of Progress: This stepwise, gradual, left-to-right imagery of human evolution long precedes Zallinger. Benjamin Waterhouse Hawkins drew standing skeletons of modern apes in Thomas Henry Huxley, *Evidence as to Man's Place in Nature* (London: Williams & Norgate, 1863). Such imagery also appeared in William K. Gregory, "The Upright Posture of Man: A Review of Its Origin and Evolution," *Proceedings of the American Philosophical Society* 67, no. 4 (1928), 339–377. It appears again on the inside cover of Raymond Dart, *Adventures with the Missing Link* (New York: Harper & Brothers, 1959).

73   our ancestors never did walk on their knuckles: C. Owen Lovejoy, Gen Suwa, Scott W. Simpson, Jay H. Matternes, and Tim D. White, "The Great Divides: *Ardipithecus ramidus* Reveals the Postcrania of Our Last Common Ancestors with African Apes," *Science* 326, no. 5949 (2009), 73–106. Tim D. White, C. Owen Lovejoy, Berhane Asfaw, Joshua P. Carlson, and Gen Suwa, "Neither Chimpanzee Nor Human, *Ardipithecus* Reveals the Surprising Ancestry of Both," *Proceedings of the National Academy of Sciences* 112, no. 16 (2015), 4877–4884.

74   *Morotopithecus*: The discoverer of this fascinating ape is Laura MacLatchy, my thesis advisor at the University of Michigan. See Laura MacLatchy, "The Oldest Ape," *Evolutionary Anthropology* 13 (2004), 90–103.

76   Uric acid helps convert fructose: James T. Kratzer et al., "Evolutionary History and Metabolic Insights of Ancient Mammalian Uricases," *Proceedings of*

the National Academy of Sciences 111, no. 10 (2014), 3763–3768. An important distinction here: Kratzer et al. use the uricase mutation as an explanation for how African apes were able to migrate back to equatorial Africa, whereas I'm suggesting that this mutation would have helped these apes live in modern Europe. There is also evidence that uric acid helps regulate blood pressure and keep it stable, even during times of starvation. Benjamin De Becker, Claudio Borghi, Michel Burnier, and Philippe van de Borne, "Uric Acid and Hypertension: A Focused Review and Practical Recommendations," *Journal of Hypertension* 37, no. 5 (2019), 878–883.

76   The presence of this gene in African apes: Matthew A. Carrigan et al., "Hominids Adapted to Metabolize Ethanol Long Before Human-Directed Fermentation," *Proceedings of the National Academy of Sciences* 112, no. 2 (2015), 458–463. For more on aye-aye alcohol metabolism, see Samuel R. Gochman, Michael B. Brown, and Nathaniel J. Dominy, "Alcohol Discrimination and Preferences in Two Species of Nectar-Feeding Primate," *Royal Society Open Science* 3 (2016), 160217.

81   Böhme and her team concluded: Madelaine Böhme et al., "A New Miocene Ape and Locomotion in the Ancestor of Great Apes and Humans," *Nature* 575 (2019), 489–493.

81   they remain controversial and contested: See Scott A. Williams et al., "Reevaluating Bipedalism in *Danuvius*," *Nature* 586 (2020), E1–E3. Madelaine Böhme, Nikolai Spassov, Jeremy M. DeSilva, and David R. Begun, "Reply to: Reevaluating Bipedalism in *Danuvius*," *Nature* 586 (2020), E4–E5.

81   He proposed that the best model: Dudley J. Morton, "Evolution of the Human Foot. II," *American Journal of Physical Anthropology* 7 (1924), 1052. Also see Russell H. Tuttle, "Darwin's Apes, Dental Apes, and the Descent of Man," *Current Anthropology* 15 (1974), 389–426. Russell H. Tuttle, "Evolution of Hominid Bipedalism and Prehensile Capabilities," *Philosophical Transactions of the Royal Society of London B* 292 (1981), 89–94.

81   They raise their arms: Personal communication with biologist Warren Brockelman.

82   It turns out that: Carol V. Ward, Ashley S. Hammond, J. Michael Plavcan, and David R. Begun, "A Late Miocene Partial Pelvis from Hungary," *Journal of Human Evolution* 136 (2019), 102645. Some scholars have also proposed that *Oreopithecus* was bipedal, though many have rejected this claim. A recent examination of its skeleton finds that the anatomies of the torso would have made *Oreopithecus* "certainly more capable of bipedal positional behaviors than extant great apes." See Ashley S. Hammond et al., "Insights into the Lower Torso in Late Miocene Hominoid *Oreopithecus bambolii*," *Proceedings of the National Academy of Sciences* 117, no. 1 (2020), 278–284.

83   Some molecular geneticists: For example, Kevin E. Langergraber et al., "Generation Times in Wild Chimpanzees and Gorillas Suggest Earlier Divergence Times in Great Ape and Human Evolution," *Proceedings of the National Academy of Sciences* 109, no. 39 (2012), 15716–15721.

84   more bipedal than chimpanzees: In his 2007 book, Aaron Filler hypothesizes that bipedal locomotion goes back even farther to the very beginning of the ape lineage 20 million years ago. He uses a backbone from *Morotopithecus bishopi* as evidence. However, nothing about the femur or hip joint of that

species would indicate bipedal locomotion in that taxon. Aaron G. Filler, *The Upright Ape: A New Origin of the Species* (Newburyport, MA: Weiser, 2007).

84   Orangutans sometimes do this today: Susannah K. S. Thorpe, Roger L. Holder, and Robin H. Crompton, "Origin of Human Bipedalism as an Adaptation for Locomotion on Flexible Branches," *Science* 316 (2007), 1328–1331.

85   light touch with the fingertips: John J. Jeka, and James R. Lackner, "Fingertip contact influences human postural control," *Experimental Brain Research* 79, no. 2 (1994), 495–502. Leif Johannsen et al., "Human Bipedal Instability in Tree Canopy Environments Is Reduced by 'Light Touch' Fingertip Support," *Scientific Reports* 7, no. 1 (2017), 1–12. This light touch might have also helped our ancestors find food. My Dartmouth College colleague Nate Dominy found that just as we humans squeeze fruit in the grocery store to assess ripeness, chimpanzees apply a light touch to figs to find the ones that are ready to eat. If chimpanzees, with their long fingers, can do this, then the earliest bipedal hominins, with more humanlike hand proportions, also did while they foraged for fruit as they walked in the trees. Nathaniel J. Dominy et al., "How Chimpanzees Integrate Sensory Information to Select Figs," *Interface Focus* 6 (2016).

86   Can the hypothesis: I have a significant number of colleagues whom I deeply respect who support a short-backed, knuckle-walking ape model for the body form from which bipedalism evolved. David Pilbeam, Dan Lieberman, David Strait, Scott Williams, and Cody Prang have all written in favor of this as the body plan of the last common ancestor. See, for example, David R. Pilbeam and Daniel E. Lieberman, "Reconstructing the Last Common Ancestor of Chimpanzees and Humans," in *Chimpanzees and Human Evolution*, ed. Martin N. Muller, Richard W. Wrangham, and David R. Pilbeam (Cambridge, MA: Belknap Press of Harvard University Press, 2017), 22–142. Some of the most convincing evidence for a knuckle-walking last common ancestor can be found in the wrist. Most primates have nine wrist bones in each hand, but humans and the African apes have only eight. The reason is that one of these bones, the os centrale, is fused to the scaphoid, making two bones one in gorillas, chimpanzees, bonobos, and humans. Why? It appears that this fusion helps stabilize the wrist during knuckle-walking and would argue in favor of a last common ancestor that moved in this manner. See Caley M. Orr, "Kinematics of the Anthropoid Os Centrale and the Functional Consequences of the Scaphoid-Centrale Fusion in African Apes and Humans," *Journal of Human Evolution* 114 (2018), 102–117. Thomas A. Püschel, Jordi Marcé-Nogué, Andrew T. Chamberlain, Alaster Yoxall, and William I. Sellers, "The Biomechanical Importance of the Scaphoid-Centrale Fusion During Simulated Knuckle-Walking and Its Implications for Human Locomotor Evolution," *Scientific Reports* 10, 3526 (2020), 1–10. This could also be interpreted as a random fusion of wrist bones that was selectively neutral in an arboreal ape that predisposed gorillas and chimpanzees to this form of locomotion later in their evolutionary history. While I currently favor a long-backed, arboreal bipedal origin model, it will be fascinating to see how this debate unfolds and is informed by new fossils in the coming decades.

86   followed the receding forests: A fossil discovery that I do not discuss in the book because I don't know what to make of it yet is the nearly 6-million-year-

old bipedal footprint site reported from the island of Crete. These footprints remain controversial, but if they are verified, then bipedal apes continued to live in European refugia even after the last common ancestor of humans and the African apes inhabited Africa. Gerard D. Gierliński et al., "Possible Hominin Footprints from the Late Miocene (c. 5.7 Ma) of Crete?" *Proceedings of the Geologists' Association* 128, no. 5–6 (2017), 697–710.

PART II: BECOMING HUMAN

87   "*Homo sapiens* didn't invent": Erling Kagge, *Walking: One Step at a Time* (New York: Pantheon, 2019), 157.

CHAPTER 6: ANCIENT FOOTPRINTS

89   "There is charm in footing slow": John Keats, Harry Buxton Forman, and Horace Elisha Scudder, *The Complete Poetical Works of John Keats* (Boston: Houghton Mifflin, 1899), 246.

89   called Laetoli: The Maasai people call the area Olaetole.

91   A bipedal child's soft, bare foot: Barefoot people develop thick calluses under their feet, which help protect the feet without sacrificing foot sensitivity. But the thorn was embedded in the middle of the child's arch, an area that would not develop a protective callus. For more on callus formation, see Nicholas B. Holowka et al., "Foot Callus Thickness Does Not Trade Off Protection for Tactile Sensitivity During Walking," *Nature* 571 (2019), 261–264.

92   There were also small, odd dimples: Peg van Andel, a middle-school science teacher in Boxborough, Massachusetts, began researching a children's book on the Laetoli footprints and interviewed Andrew Hill before he died. In these interview notes, Hill mentioned these raindrop impressions and their connection to Lyell's *Principles of Geology*.

93   Animals walked through it for a few days: See Mary D. Leakey and Richard L. Hay, "Pliocene Footprints in the Laetolil Beds at Laetoli, Northern Tanzania," *Nature* 278 (1979), 317–323. Mary Leakey, "Footprints in the Ashes of Time," *National Geographic* 155, no. 4 (1979), 446–457. Michael H. Day and E. H. Wickens, "Laetoli Pliocene Hominid Footprints and Bipedalism," *Nature* 286 (1980), 385–387. Mary D. Leakey and Jack M. Harris, eds., *Laetoli: A Pliocene Site in Northern Tanzania* (Oxford: Oxford University Press, 1987). Tim D. White and Gen Suwa, "Hominid Footprints at Laetoli: Facts and Interpretations," *American Journal of Physical Anthropology* 72 (1987), 485–514. Neville Agnew and Martha Demas, "Preserving the Laetoli Footprints," *Scientific American* (1998), 44–55. The commonly held idea that the nearby Sadiman volcano is the source of the Laetoli ash has recently been challenged, making the source of the ash currently unknown. See Anatoly N. Zaitsev et al., "Stratigraphy, Minerology, and Geochemistry of the Upper Laetolil Tuffs Including a New Tuff 7 Site with Footprints of *Australopithecus afarensis*, Laetoli, Tanzania," *Journal of African Earth Sciences* 158 (2019), 103561.

93   In September, they did: Details from Mary Leakey, *Disclosing the Past: An Autobiography* (New York: Doubleday, 1984). Virginia Morell, *Ancestral Passions: The Leakey Family and the Quest for Humankind's Beginnings* (New York: Simon & Schuster, 1995).

94   She sent Ndibo Mbuika: Other prominent researchers involved in the discovery and excavation of the G-trails were Tim White, Ron Clarke, Michael Day, and Louise Robbins.

94   It appears that three, perhaps even four: See Matthew R. Bennett, Sally C. Reynolds, Sarita Amy Morse, and Marcin Budka, "Laetoli's Lost Tracks: 3D Generated Mean Shape and Missing Footprints," *Scientific Reports* 6 (2016), 21916. Charles Musiba has proposed that there may be four individuals making the Laetoli G-trail.

95   It appears, too, that the Laetoli footprints: See Kevin G. Hatala, Brigitte Demes, and Brian G. Richmond, "Laetoli Footprints Reveal Bipedal Gait Biomechanics Different from Those of Modern Humans and Chimpanzees," *Proceedings of the Royal Society B: Biological Sciences* 283, no. 1836 (2016), 20160235.

99   They may have been made: In 2021, our team completed a thorough analysis of the fossil footprints and concluded that they were made by a second species of hominin walking at Laetoli. I discuss this more in Chapter 7.

99   While the students cleaned the prints: We could have made a bipedalism playlist consisting of "Walk of Life" (Dire Straits), "Love Walks In" (Van Halen), "Walking on a Thin Line" (Huey Lewis), "Walking on Sunshine" (Katrina and the Waves), and "Walk This Way" (Run-DMC version, of course).

101   a new species, *Homo habilis*: Louis S. B. Leakey, Phillip V. Tobias, and John R. Napier, "A New Species of the Genus *Homo* from Olduvai Gorge," *Nature* 202, no. 4927 (1964), 7–9.

102   deposited 3.3 million years ago: Sonia Harmand et al., "3.3-Million-Year-Old Stone Tools from Lomekwi 3, West Turkana, Kenya," *Nature* 521 (2015), 310–315.

102   partial skeleton of an *Australopithecus* toddler: Zeresenay Alemseged et al., "A Juvenile Early Hominin Skeleton from Dikika, Ethiopia," *Nature* 443 (2006), 296–301. Jeremy M. DeSilva, Corey M. Gill, Thomas C. Prang, Miriam A. Bredella, and Zeresenay Alemseged, "A Nearly Complete Foot from Dikika, Ethiopia, and Its Implications for the Ontogeny and Function of *Australopithecus afarensis*," *Science Advances* 4, no. 7 (2018), eaar7723.

102   deliberately cut by sharp rocks: Shannon P. McPherron et al., "Evidence for Stone-Tool-Assisted Consumption of Animal Tissues Before 3.39 Million Years Ago at Dikika, Ethiopia," *Nature* 466 (2010), 857–860.

103   "I feel that scientists holding to this definition": Baroness Jane Van Lawick-Goodall, *My Friends the Wild Chimpanzees* (Washington, DC: National Geographic Society, 1967), 32.

105   Evidence from the genetics: David L. Reed, Jessica E. Light, Julie M. Allen, and Jeremy J. Kirchman, "Pair of Lice Lost or Parasites Regained: The Evolutionary History of Anthropoid Lice," *BMC Biology* 5, no. 7 (2007). Note the title of this paper.

106   Maybe an older child held the baby: See Rebecca Sear and David Coall, "How Much Does Family Matter? Cooperative Breeding and the Demographic Transition," *Population and Development Review* 37, no. s1 (2011), 81–112.

106   small acts of parenting by others: This idea is known as the cooperative breeding hypothesis, and it was developed by Sarah Hrdy in her extraor-

dinary book *Mothers and Others: The Evolutionary Origins of Mutual Understanding* (Cambridge, MA: Belknap Press, 2009).

106 collectively raise our children: Jeremy M. DeSilva, "A Shift Toward Birthing Relatively Large Infants Early in Human Evolution," *Proceedings of the National Academy of Sciences* 108, no. 3 (2011), 1022–1027. One prediction of the hypothesis that *Australopithecus* was collectively parenting their young is weaning age. In great apes, mothers nurse their babies for over four years. Orangutan young do not wean until they are over seven years old. Humans in hunter-gatherer communities, in contrast, nurse between one and four years. We can afford to wean early, in part, because there are other members of the group able and willing to share food. Recent analyses of the isotopes in *Australopithecus* infant teeth reveal that they, too, weaned early. This is independent evidence for cooperative raising of the young in our early ancestors. Théo Tacail et al., "Calcium Isotopic Patterns in Enamel Reflect Different Nursing Behaviors Among South African Early Hominins," *Science Advances* 5 (2019), eaax3250. Renaud Joannes-Boyau et al., "Elemental Signatures of *Australopithecus africanus* Teeth Reveal Seasonal Dietary Stress," *Nature* 572 (2019), 112–116.

108 They would have eaten whatever: Some of this evidence comes from carbon isotopes, which show a wide range of values in *Australopithecus afarensis*. Jonathan G. Wynn, "Diet of *Australopithecus afarensis* from the Pliocene Hadar Formation, Ethiopia," *Proceedings of the National Academy of Sciences* 110, no. 26 (2013), 10495–10500.

108 This shift toward a generalized diet: See Daniel Lieberman, *The Story of the Human Body: Evolution, Health, and Disease* (New York: Vintage, 2013).

108 Today, baboons and chimpanzees: Jane Goodall, *The Chimpanzees of Gombe: Patterns of Behavior* (Cambridge, MA: Harvard University Press, 1986), 555–557. Jane Goodall, "Tool-Using and Aimed Throwing in a Community of Free-Living Chimpanzees," *Nature* 201 (1964), 1264–1266. William J. Hamilton, Ruth E. Buskirk, and William H. Buskirk, "Defensive Stoning by Baboons," *Nature* 256 (1975), 488–489. Martin Pickford, "Matters Arising: Defensive Stoning by Baboons (Reply)," *Nature* 258 (1975), 549–550.

109 Recently, paleoanthropologist Yohannes Haile-Selassie: Yohannes Haile-Selassie, Stephanie M. Melillo, Antonino Vazzana, Stefano Benazzi, and Timothy M. Ryan, "A 3.8-Million-Year-Old Hominin Cranium from Woranso-Mille, Ethiopia," *Nature* 573 (2019), 214–219.

109 That's still only a third the brain: William H. Kimbel, Yoel Rak, and Donald C. Johanson, *The Skull of* Australopithecus afarensis (Oxford: Oxford University Press, 2004).

109 Your brain is only 2 percent: Even more is required when a child's brain is growing—over 40 percent of the body's energy. Christopher W. Kuzawa et al., "Metabolic Costs and Evolutionary Implications of Human Brain Development," *Proceedings of the National Academy of Sciences* 111, no. 36 (2014), 13010–13015.

110 In a grassland environment: See Herman Pontzer, "Economy and Endurance in Human Evolution," *Current Biology* 27 (2017), R613–R621.

111 At this age, the child had grown: Dikika paper on brain growth. Philipp Gunz et al., "*Australopithecus afarensis* Endocasts Suggest Apelike Brain Organization and Prolonged Brain Growth," *Science Advances* 6 (2020), eaaz4729. In

fact, Smith was able to age the Dikika Child to the *day*. She was 861 days old when she died. As with many other studies discussed in this book, this was a large team effort. The scans Tanya Smith examined were collected in collaboration with Paul Tafforeau and Adeline LeCabec. Philipp Gunz reconstructed the brain of the child and, of course, Zeray Alemseged found the fossil in the first place.

CHAPTER 7: MANY WAYS TO WALK A MILE

113   "There is more than one way": Ann Gibbons, "Skeletons Present an Exquisite Paleo-Puzzle," *Science* 333 (2011), 1370–1372. Bruce Latimer, personal communication.

114   closest thing we have in paleoanthropology to Indiana Jones: What I mean by this is that Lee Berger is an explorer and an adventurer who inspires future generations of researchers by popularizing our science. Thus, Berger is Indiana Jones–like in all of the good ways, not in the philandering and pilfering ways Harrison Ford's character also embodied.

115   Lee Berger downloaded: More details of this gripping story can be found in Berger's two books on the subject. Lee Berger and Marc Aronson, *The Skull in the Rock: How a Scientist, a Boy, and Google Earth Opened a New Window on Human Origins* (Washington, DC: National Geographic Children's Books, 2012). Lee Berger and John Hawks, *Almost Human: The Astonishing Tale of* Homo naledi *and the Discovery That Changed Our Human Story* (Washington, DC: National Geographic, 2017).

116   the probable cause of death: Ericka N. L'Abbé et al., "Evidence of Fatal Skeletal Injuries on Malapa Hominins 1 and 2," *Scientific Reports* 5, no. 15120 (2015).

117   within a 3,000-year window 1.977 million years ago: Robyn Pickering et al., "*Australopithecus sediba* at 1.977 Ma and Implications for the Origins of the Genus *Homo*," *Science* 333, no. 6048 (2011), 1421–1423.

117   They named it *sediba*: Lee Berger et al., "*Australopithecus sediba*: A New Species of *Homo*-Like Australopith from South Africa," *Science* 328, no. 5975 (2010), 195–204.

118   Berger and Zipfel had sent plastic casts: This needs some explanation for readers who are wondering why I would be surprised at the anatomy of the casts after having seen the originals under the black cloth in Berger's lab. The original fossil foot and ankle bones (tibia, talus, and calcaneus) are still articulated and held together by matrix. They were micro-CT-scanned by Kristian Carlson and digitally pulled apart after hours of tedious computer work. Carlson then 3D-printed the digital renderings of the isolated foot bones. Those were what Berger and Zipfel had sent me in the spring of 2010.

120   Noting differences in their pelvises: John T. Robinson, *Early Hominid Posture and Locomotion* (Chicago: University of Chicago Press, 1972). Robinson also classified *africanus* as *Homo*. If ever adopted, this would wreak havoc on the names of hominins since the type species for *Australopithecus* is *africanus*.

120   Thirty years later, American Museum: William E. H. Harcourt-Smith and Leslie C. Aiello, "Fossils, Feet, and the Evolution of Human Bipedal Locomotion," *Journal of Anatomy* 204, no. 5 (2004), 403–416.

121   Zipfel and I published our findings: Bernhard Zipfel et al., "The Foot and

Ankle of *Australopithecus sediba*," *Science* 333, no. 6048 (2011), 1417–1420. As with other studies described in this book, this was a team effort, and significant contributions were made by Robert Kidd, Kristian Carlson, Steve Churchill, and Lee Berger.

123 Holt, Zipfel, and I tested our hypothesis: Jeremy M. DeSilva et al., "The Lower Limb and Mechanics of Walking in *Australopithecus sediba*," *Science* 340, no. 6129 (2013), 1232999.

123 We took MRI scans of forty humans: Jeremy M. DeSilva et al., "Midtarsal Break Variation in Modern Humans: Functional Causes, Skeletal Correlates, and Paleontological Implications," *American Journal of Physical Anthropology* 156, no. 4 (2015), 543–552.

123 rig the joints and make it walk: Amey Y. Zhang and Jeremy M. DeSilva, "Computer Animation of the Walking Mechanics of *Australopithecus sediba*," *PaleoAnthropology* (2018), 423–432. Sally Le Page tweet of *sediba* walking: https://twitter.com/sallylepage/status/1088364360857198598.

124 "Monty Python's 'Ministry of Silly Walks' sketch": William H. Kimbel, "Hesitation on Hominin History," *Nature* 497 (2013), 573–574. For "Ministry of Silly Walks" sketch, see: https://www.dailymotion.com/video/x2hwqki. For a brilliant paper analyzing the gaits of the minister and Mr. Pudley, see Erin E. Butler and Nathaniel J. Dominy, "Peer Review at the Ministry of Silly Walks," *Gait & Posture* (February 26, 2020).

125 Marion Bamford, the director of: Marion Bamford et al., "Botanical Remains from a Coprolite from the Pleistocene Hominin Site of Malapa, Sterkfontein Valley, South Africa," *Palaeontologica Africana* 45 (2010), 23–28.

125 With its long arms and shrugged shoulders: The long arms probably do not need explaining as an adaptation for climbing in the trees, but the shrugged shoulders might. Kevin Hunt proposed that narrow, shrugged shoulders would help balance the center of mass of an arm-hanging ape. Kevin D. Hunt, "The Postural Feeding Hypothesis: An Ecological Model for the Evolution of Bipedalism," *South African Journal of Science* 92 (1996), 77–90.

125 it relied heavily on food from forests: Amanda G. Henry et al., "The Diet of *Australopithecus sediba*," *Nature* 487 (2012), 90–93.

128 He has since discovered: Yohannes Haile-Selassie et al., "New Species from Ethiopia Further Expands Middle Pliocene Hominin Diversity," *Nature* 521 (2015), 483–488.

129 "You found another Ardi!": Latimer quotes from John Mangels, "New Human Ancestor Walked and Climbed 3.4 Million Years Ago in Lucy's Time, Cleveland Team Finds (Video)," *Cleveland Plain Dealer* (March 28, 2012), https://www.cleveland.com/science/2012/03/new_human_ancestor_walked_and.html.

130 Another hominin, walking in a different: Yohannes Haile-Selassie et al., "A New Hominin Foot from Ethiopia Shows Multiple Pliocene Bipedal Adaptations," *Nature* 483 (2012), 565–569. But it is important to note that Haile-Selassie has not directly attributed the Burtele foot to *Australopithecus deyiremeda*. It could be from a third, as-yet-unnamed hominin.

130 In fact, in 2021 we completed: Ellison McNutt et al., "Footprint Evidence of Early Hominin Locomotor Diversity at Laetoli, Tanzania," *Nature* 600, no. 7889 (2021), 468–471.

CHAPTER 8: HOMININS ON THE MOVE

131   "There was nowhere to go": Jack Kerouac, *On the Road* (New York: Viking Press, 1957), 26.

131   What was a rhinoceros doing: In the thirteenth century, Marco Polo traveled over 7,500 miles from his home in Italy to China along the Silk Road. Maps of his journey take him past Dmanisi, though no evidence is known that he stopped there. Many travelers did, however, as Dmanisi became an important part of the trade route between Europe and Asia and was eventually absorbed into the Mongol Empire. Polo carried on, and when he reached the island of Java, he reports seeing a unicorn. He wrote, "There are wild elephants in the country, and numerous unicorns, which are very nearly as big. They have hair like that of a buffalo, feet like those of an elephant, and a horn in the middle of the forehead, which is black and very thick. They do no mischief, however, with the horn, but with the tongue alone; for this is covered all over with long and strong prickles [and when savage with any one they crush him under their knees and then rasp him with their tongue]. The head resembles that of a wild boar, and they carry it ever bent towards the ground. They delight much to abide in mire and mud. 'Tis a passing ugly beast to look upon, and is not in the least like that which our stories tell of as being caught in the lap of a virgin; in fact, 'tis altogether different from what we fancied." What Polo describes in his *Travels* is, of course, a rhinoceros.

132   in 1991, they found a hominin jaw: Leo Gabunia and Abesalom Vekua, "A Plio-Pleistocene Hominid from Dmanisi, East Georgia, Caucasus," *Nature* 373 (1995), 509–512.

132   In 2018, Zhaoyu Zhu: Zhaoyu Zhu et al., "Hominin Occupation of the Chinese Loess Plateau Since About 2.1 Million Years Ago," *Nature* 559 (2018), 608–612.

134   Three years after I met him: Fred Spoor et al., "Implications of New Early *Homo* Fossils from Ileret, East of Lake Turkana, Kenya," *Nature* 448 (2007), 688–691. Fredrick Manthi's current title is Head of Department of Earth Sciences at the National Museums of Kenya.

135   In their book *The Wisdom*: Alan Walker and Pat Shipman, *The Wisdom of the Bones: In Search of Human Origins* (New York: Vintage, 1997).

135   "Lord knows how he saw it": Walker and Shipman, *The Wisdom of the Bones*, 12.

136   Scientists calculate that the Nariokotome: There is uncertainty about this. It starts with how old the Nariokotome child was when he died. Chronological age estimates range from 7.6 to 8.8 years old to as high as 15 years old, though most scholars cite the younger age ranges determined from state-of-the-art analysis of tooth development. Height at death for the child ranges from four feet eight inches to five feet three inches depending on what technique is employed. Then, there is the question of whether *Homo erectus* had an adolescent growth spurt, or if this evolved more recently. The adult size of Nariokotome is calculated to be somewhere between five feet four inches and over six feet. See Ronda R. Graves, Amy C. Lupo, Robert C. McCarthy, Daniel J. Wescott, and Deborah L. Cunningham, "Just How Strapping Was KNM-WT 15000?" *Journal of Human Evolution* 59, no. 5 (2010), 542–554. Chris Ruff and Alan Walker, "Body Size and Body Shape" in *The Narioko-*

tome Homo erectus *Skeleton*, ed. Alan Walker and Richard Leakey (Cambridge, MA: Harvard University Press, 1993), 234–265.

136 Northwestern University anthropologist Chris Kuzawa: Christopher W. Kuzawa et al., "Metabolic Costs and Evolutionary Implications of Human Brain Development," *Proceedings of the National Academy of Sciences* 111, no. 36 (2014), 13010–13015.

136 Its large right femur is the size: Henry M. McHenry, "Femoral Lengths and Stature in Plio-Pleistocene Hominids," *American Journal of Physical Anthropology* 85 (1991), 149–158. A slightly shorter estimate of height for KNM-ER 1808 of five feet eight inches was obtained by Manuel Will and Jay T. Stock, "Spatial and Temporal Variation of Body Size Among Early *Homo*," *Journal of Human Evolution* 82 (2015), 15–33. On the basis of footprint size, a team estimated that a group of *Homo erectus* ranged in height from five feet to just over six feet. Heather L. Dingwall, Kevin G. Hatala, Roshna E. Wunderlich, and Brian G. Richmond, "Hominin Stature, Body Mass, and Walking Speed Estimates Based on 1.5-Million-Year-Old Fossil Footprints at Ileret, Kenya," *Journal of Human Evolution* 64, no. 6 (2013), 556–568.

137 In 2009, a team of researchers: Matthew R. Bennett et al., "Early Hominin Foot Morphology Based on 1.5-Million-Year-Old-Footprints from Ileret, Kenya," *Science* 323, no. 5918 (2009), 1197–1201. Kevin G. Hatala et al., "Footprints Reveal Direct Evidence of Group Behavior and Locomotion in *Homo erectus*," *Scientific Reports* 6 (2016), 28766.

137 especially when they ran: Dennis M. Bramble and Daniel E. Lieberman, "Endurance Running and the Evolution of *Homo*," *Nature* 432 (2004), 345–352.

137 In ecosystems throughout the world: Chris Carbone, Guy Cowlishaw, Nick J. B. Isaac, and J. Marcus Rowcliffe, "How Far Do Animals Go? Determinants of Day Range in Mammals," *American Naturalist* 165, no. 2 (2005), 290–297.

138 Either way, 2.1 million: One should wonder what role the environment played in creating conditions conducive for hominin migration into Eurasia. There is evidence for global drying and cooling—and subsequent expansion of grassland habitats—caused, in part, by altered ocean currents resulting from the physical separation of the Atlantic and Pacific Oceans thanks to the closing of the Isthmus of Panama 2.8 million years ago. See Aaron O'Dea et al., "Formation of the Isthmus of Panama," *Science Advances* 2, no. 8 (2016), e1600883. Steven M. Stanley, *Children of the Ice Age: How a Global Catastrophe Allowed Humans to Evolve* (New York: Crown, 1996).

139 During ice ages: Eight periods of glaciation are known from the last three-quarters of a million years. EPICA community members, "Eight Glacial Cycles from an Antarctic Ice Core," *Nature* 429 (2004), 623–628.

139 The researchers called the fossil: Isidro Toro-Moyano et al., "The Oldest Human Fossil in Europe, from Orce (Spain)," *Journal of Human Evolution* 65, no. 1 (2013), 1–9. Eudald Carbonell et al., "The First Hominin of Europe," *Nature* 452 (2008), 465–469. José María Bermúdez de Castro et al., "A Hominid from the Lower Pleistocene of Atapuerca, Spain: Possible Ancestor to Neandertals and Modern Humans," *Science* 276, no. 5317 (1997), 1392–1395.

140 The first, formulated by anthropologists: Leslie C. Aiello and Peter Wheeler, "The Expensive-Tissue Hypothesis: The Brain and the Digestive System in Human and Primate Evolution," *Current Anthropology* 36, no. 2 (1995), 199–221.

140   More recently, Richard Wrangham: Richard Wrangham, *Catching Fire: How Cooking Made Us Human* (New York: Basic Books, 2009). The only problem with this elegant hypothesis is timing. The earliest evidence for controlled fire is 1.5 million years old. But brain increase is detectable in the fossil record starting at least 2 million years ago. Either controlled fire is older than we currently have evidence for, or cooking cannot explain the initial increase in brain size in early *Homo*. Even if the latter ends up being supported by paleontological and archaeological evidence, controlled fire and cooking almost certainly maintained and perhaps even accelerated brain growth in Pleistocene *Homo*.

140   fire is a predator deterrent: See Richard Wrangham and Rachel Carmody, "Human Adaptation to the Control of Fire," *Evolutionary Anthropology* 19 (2010), 187–199.

141   But animals that walk on two feet: See Dennis M. Bramble and David R. Carrier, "Running and Breathing in Mammals," *Science* 219, no. 4582 (1983), 251–256. Robert R. Provine, "Laughter as an Approach to Vocal Evolution," *Psychonomic Bulletin & Review* 23 (2017), 238–244.

141   In a seated position: See Morgan L. Gustison, Aliza le Rouz, and Thore J. Bergman, "Derived Vocalizations of Geladas (*Theropithecus gelada*) and the Evolution of Vocal Complexity in Primates," *Philosophical Transactions of the Royal Society B* 367, no. 1597 (2012). I wonder how far this relationship between vocalization and locomotion extends. Birds, for instance, have an extraordinary vocal repertoire. Aquatic animals such as whales and dolphins, their chest muscles buoyed by water, also have complex communication systems.

142   Even in our children: In both American and Chinese children, first steps and first words are correlated, independent of the age in which they occurred. Minxuan He, Eric A. Walle, and Joseph J. Campos, "A Cross-National Investigation of the Relationship Between Infant Walking and Language Development," *Infancy* 20, no. 3 (2015), 283–305.

142   asymmetry in the Broca region: See review in Amélie Beaudet, "The Emergence of Language in the Hominin Lineage: Perspectives from Fossil Endocasts," *Frontiers in Human Neuroscience* 11 (2017), 427. Dean Falk, "Interpreting Sulci on Hominin Endocasts: Old Hypotheses and New Findings," *Frontiers in Human Neuroscience* 8 (2014), 134. Certainly, this was the case for early *Homo* as evidenced by KNM-ER 1470. Dean Falk, "Cerebral Cortices of East African Early Hominids," *Science* 221, no. 4615 (1983), 1072–1074.

143   Half million-year-old fossils: See Ignacio Martínez et al., "Auditory Capacities in Middle Pleistocene Humans from the Sierra de Atapuerca in Spain," *Proceedings of the National Academy of Sciences* 101, no. 27 (2004), 9976–9981. Ignacio Martínez et al., "Communicative Capacities in Middle Pleistocene Humans from the Sierra de Atapuerca in Spain," *Quaternary International* 295 (2013), 94–101. Ignacio Martínez et al., "Human Hyoid Bones from the Middle Pleistocene Site of the Sima de los Huesos (Sierra de Atapuerca, Spain)," *Journal of Human Evolution* 54, no. 1 (2008), 118–124. Johannes Krause et al., "The Derived *FOXP2* Variant of Modern Humans Was Shared with Neandertals," *Current Biology* 17, no. 21 (2007), 1908–1912. See also Elizabeth G. Atkinson et al., "No Evidence for Recent

Selection of *FOXP2* Among Diverse Human Populations," *Cell* 174, no. 6 (2018), 1424–1435.

143   Around 800,000: Nick Ashton et al., "Hominin Footprints from Early Pleistocene Deposits at Happisburgh, UK," *PLOS ONE* 9, no. 2 (2014), e88329.

144   In 2019, scientists from the National: Jérémy Duveau, Gilles Berillon, Christine Verna, Gilles Laisné, and Dominique Cliquet, "The Composition of a Neandertal Social Group Revealed by the Hominin Footprints at Le Rozel (Normandy, France)," *Proceedings of the National Academy of Sciences* 116, no. 39 (2019), 19409–19414.

144   DNA extracted from tiny scraps: David Reich et al., "Genetic History of an Archaic Hominin Group from Denisova Cave in Siberia," *Nature* 468, no. 7327 (2010), 1053–1060. Fahu Chen et al., "A Late Middle Pleistocene Denisovan Mandible from the Tibetan Plateau," *Nature* 569 (2019), 409–412.

CHAPTER 9: MIGRATION TO MIDDLE EARTH

145   "Not all who wander are lost": From the poem "All That Is Gold Does Not Glitter" in J. R. R. Tolkien, *Lord of the Rings: The Fellowship of the Ring* (London: George Allen & Unwin, 1954).

146   once thick enough to cover the 6,288-foot Mt. Washington: Though the ice at the top of the mountain would have been thin since the rocks at the summit are not scoured.

147   In 2019, for example, a study published: Eva K. F. Chan et al., "Human Origins in a Southern African Palaeo-Wetland and First Migrations," *Nature* 575 (2019), 185–189.

148   Recent studies examining entire: Carina M. Schlebusch et al., "Southern African Ancient Genomes Estimate Modern Human Divergence to 350,000 to 260,000 Years Ago," *Science* 358, no. 6363 (2017), 652–655.

148   Smithsonian Institution scientists: Alison S. Brooks et al., "Long-Distance Stone Transport and Pigment Use in the Earliest Middle Stone Age," *Science* 360, no. 6384 (2018), 90–94.

149   In 2019, Katerina Harvati: Katerina Harvati et al., "Apidima Cave Fossils Provide Earliest Evidence of *Homo sapiens* in Eurasia," *Nature* 571 (2019), 500–504. Israel Hershkovitz et al., "The Earliest Modern Humans Outside Africa," *Science* 359, no. 6374 (2018), 456–459.

149   We know from DNA miraculously: Richard E. Green et al., "Analysis of One Million Base Pairs of Neanderthal DNA," *Nature* 444 (2006), 330–336. Lu Chen, Aaron B. Wolf, Wenqing Fu, Liming Li, and Joshua M. Akey, "Identifying and Interpreting Apparent Neanderthal Ancestry in African Individuals," *Cell* 180, no. 4 (2020), 677–687.

150   We were on the Australian: Chris Clarkson et al., "Human Occupation of Northern Australia by 65,000 Years Ago," *Nature* 547 (2017), 306–310.

150   By 20,000 years ago: Steve Webb, Matthew L. Cupper, and Richard Robbins, "Pleistocene Human Footprints from the Willandra Lakes, Southeastern Australia," *Journal of Human Evolution* 50, no. 4 (2006), 405–413.

151   The Fort Rock cave sandals: See references in Janna T. Kuttruff, S. Gail DeHart, and Michael J. O'Brien, "7500 Years of Prehistoric Footwear from Arnold Research Cave," *Science* 281, no. 5373 (1998), 72–75.

151   Erik Trinkaus of Washington University: Erik Trinkaus, "Anatomical Evidence for the Antiquity of Human Footwear Use," *Journal of Archaeological Science* 32, no. 10 (2005), 1515–1526. Erik Trinkaus and Hong Shang, "Anatomical Evidence for the Antiquity of Human Footwear: Tianyuan and Sunghir," *Journal of Archaeological Science* 35, no. 7 (2008), 1928–1933.

151   Thirteen thousand years ago: Duncan McLaren et al., "Terminal Pleistocene Epoch Human Footprints from the Pacific Coast of Canada," *PLOS ONE* 13, no. 3 (2018), e0193522. Karen Moreno et al., "A Late Pleistocene Human Footprint from the Pilauco Archaeological Site, Northern Patagonia, Chile," *PLOS ONE* 14, no. 4 (2019), e0213572.

152   On the morning of September: Some of this information came from Paige Madison, "Floresiensis Family: Legacy & Discovery at Liang Bua," April 26, 2018, http://fossilhistorypaige.com/2018/04/lunch-liang-bua.

153   Researchers declared it to be: Peter Brown et al., "A New Small-Bodied Hominin from the Late Pleistocene of Flores, Indonesia," *Nature* 431 (2004), 1055–1061.

153   It shared foot and hand anatomies: William L. Jungers et al., "The Foot of *Homo floresiensis*," *Nature* 459 (2009), 81–84.

154   The discoverers named a new: Florent Détroit et al., "A New Species of *Homo* from the Late Pleistocene of the Philippines," *Nature* 568 (2019), 181–186.

156   When word reached Berger: Details spelled out in Lee Berger and John Hawks, *Almost Human: The Astonishing Tale of* Homo naledi *and the Discovery That Changed Our Human Story* (Washington, DC: National Geographic, 2017).

158   We named it *Homo naledi*: Lee R. Berger et al., "*Homo naledi*, a New Species of the Genus *Homo* from the Dinaledi Chamber, South Africa," *eLife* 4 (2015), e09560. *Homo naledi* fossils have been found in a second chamber in the Rising Star cave system: John Hawks et al., "New Fossil Remains of *Homo naledi* from the Lesedi Chamber, South Africa," *eLife* 6 (2017), e24232. As with the *Australopithecus sediba* fossils from Malapa Cave, South Africa, a large number of *Homo naledi* fossils have been surface-scanned, and digital models of them are available at www.morphosource.org.

158   The bones are only 260,000: Paul H. G. M. Dirks et al., "The Age of *Homo naledi* and Associated Sediments in the Rising Star Cave, South Africa," *eLife* 6 (2017), e24231.

159   We don't know why: Ian Tattersall attributes the survival of *Homo sapiens* over other hominins to our symbolic behavior. See Ian Tattersall, *Masters of the Planet* (New York: Palgrave Macmillan, 2012). Pat Shipman proposes that the domestication of the dog gave humans an advantage, particularly over Neandertals. See Pat Shipman, *The Invaders: How Humans and Their Dogs Drove Neanderthals to Extinction* (Cambridge, MA: Belknap Press of Harvard University Press, 2015).

PART III: WALK OF LIFE

161   "Afoot and light-hearted": Walt Whitman, "Song of the Open Road," in *Leaves of Grass* (Self-published, 1855).

CHAPTER 10: BABY STEPS

165 But humans are different: See Wenda Trevathan and Karen Rosenberg, eds., *Costly and Cute: Helpless Infants and Human Evolution* (Santa Fe: University of New Mexico Press, published in association with School for Advanced Research Press, 2016).

165 can mimic some facial expressions: See Andrew N. Meltzoff and M. Keith Moore, "Imitation of Facial and Manual Gestures by Human Neonates," *Science* 198, no. 4312 (1977), 75–78.

165 a video taken moments after a birth: See critique of media sensationalism around this video by Dr. Jen Gunter on her blog, "A Newborn Baby in Brazil Didn't Walk, Journalists Made a Story of a Normal Reflex. That's Wrong," May 30, 2017, https://drjengunter.com/2017/05/30/a-newborn-baby-in-brazil -didnt-walk-journalists-made-a-story-of-a-normal-reflex-thats-wrong.

165 Albrecht Peiper, a German pediatrician: Albrecht Peiper, *Cerebral Function in Infancy and Childhood* (New York: Consultants Bureau, 1963).

166 Alessandra Piontelli, who studies fetal: Alessandra Piontelli, *Development of Normal Fetal Movements: The First 25 Weeks of Gestation* (Milan: Springer-Verlag Italia, 2010).

166 At first, Nadia Dominici, a neuroscientist: Nadia Dominici et al., "Locomotor Primitives in Newborn Babies and Their Development," *Science* 334, no. 6058 (2011), 997–999.

166 About fifty years ago, Philip Roman Zelazo: Philip Roman Zelazo, Nancy Ann Zelazo, and Sarah Kolb, "'Walking' in the Newborn," *Science* 176 (1972), 314–315.

166 little chubby legs: In fact, it appears that these chubby legs may play a role in delaying the transition of a "step reflex" into actual walking for—on average—a year. See Esther Thelen and Donna M. Fisher, "Newborn Stepping: An Explanation for a 'Disappearing' Reflex," *Developmental Psychology* 18, no. 5 (1982), 760–775.

167 a child walks independently: There are strict criteria for what counts as walking independently. Some define walking onset as taking five consecutive steps. Others define it as being able to walk ten feet without stopping or falling.

168 between thirteen and fifteen months: Apparently, however, Gesell only collected data on babies from German heritage and excluded babies of single parents. Such exclusion makes extrapolating his data to a population average a deeply flawed exercise.

168 Things shifted again in 1992: Beth Ellen Davis, Rachel Y. Moon, Hari C. Sachs, and Mary C. Ottolini, "Effects of Sleep Position on Infant Motor Development," *Pediatrics* 102, no. 5 (1998), 1135–1140.

168 As anthropologists Kate Clancy: Kathryn B. H. Clancy and Jenny L. Davis, "Soylent Is People, and WEIRD Is White: Biological Anthropology, Whiteness, and the Limits of the WEIRD," *Annual Review of Anthropology* 48 (2019), 169–186.

169 Botflies deposit their larvae: Kim Hill and A. Magdalena Hurtado, *Ache Life History* (New York: Routledge, 1996), 153–154.

169 "An infant or small child": Hill and Hurtado, *Ache Life History*, 154.

169 Anthropologists Hillard Kaplan and: Hillard Kaplan and Heather Dove,

"Infant Development Among the Ache of Eastern Paraguay," *Developmental Psychology* 23, no. 2 (1987), 190–198.

170   In parts of northern China: See references in Karen Adolph and Scott R. Robinson, "The Road to Walking: What Learning to Walk Tells Us About Development," in *Oxford Handbook of Developmental Psychology*, ed. Philip David Zelazo (Oxford: Oxford University Press, 2013). Lana B. Karasik, Karen E. Adolph, Catherine S. Tamis-LeMonda, and Marc H. Bornstein, "WEIRD Walking: Cross-Cultural Research on Motor Development," *Behavioral and Brain Sciences* 33, no. 2–3 (2010), 95–96.

170   A Swiss study of 220: Oskar G. Jenni, Aziz Chaouch, Jon Caflisch, and Valentin Rousson, "Infant Motor Milestones: Poor Predictive Value for Outcome of Healthy Children," *Acta Paediatrica* 102 (2013), e181–e184. Graham K. Murray, Peter B. Jones, Diana Kuh, and Marcus Richards, "Infant Developmental Milestones and Subsequent Cognitive Function," *Annals of Neurology* 62, no. 2 (2007), 128–136.

171   Every once in a while: Trine Flensborg-Madsen and Erik Lykke Mortensen, "Infant Developmental Milestones and Adult Intelligence: A 34-Year Follow-Up," *Early Human Development* 91, no. 7 (2015), 393–400. Akhgar Ghassabian et al., "Gross Motor Milestones and Subsequent Development," *Pediatrics* 138, no. 1 (2016), e20154372.

171   opens the door to new learning opportunities: Joseph J. Campos et al., "Travel Broadens the Mind," *Infancy* 1, no. 2 (2000), 149–219.

171   In 2015, however, researchers studying: Alex Ireland, Adrian Sayers, Kevin C. Deere, Alan Emond, and Jon H. Tobias, "Motor Competence in Early Childhood Is Positively Associated with Bone Strength in Late Adolescence," *Journal of Bone and Mineral Research* 31, no. 5 (2016), 1089–1098. This same research group found in 2017 that late walking as a child was a predictor for low bone strength in a group of sixty- to sixty-four-year-olds. Alex Ireland et al., "Later Age at Onset of Independent Walking Is Associated with Lower Bone Strength at Fracture-Prone Sites in Older Men," *Journal of Bone and Mineral Research* 32, no. 6 (2017), 1209–1217. Charlotte L. Ridgway et al., "Infant Motor Development Predicts Sports Participation at Age 14 Years: Northern Finland Birth Cohort of 1966," *PLOS ONE* 4, no. 8 (2009), e6837.

171   Muhammad Ali, when he was baby: From Jonathan Eig, *Ali: A Life* (Boston: Houghton Mifflin Harcourt, 2017), 11. James S. Hirsch, *Willie Mays: The Life, the Legend* (New York: Scribner, 2010), 13. Andrew S. Young, *Black Champions of the Gridiron* (New York: Harcourt, Brace & World, 1969). Martin Kessler, "Kalin Bennett Has Autism—and He's a Div. I Basketball Player," *Only a Game*, WBUR, June 21, 2019, https://www.wbur.org/onlyagame/2019/06/21/kent-state-kalin-bennett-basketball-autism.

172   Many children in cultures: See references in Adolph and Robinson, "The Road to Walking."

172   "Individual infants forge their": Adolph and Robinson, "The Road to Walking," 410.

173   As Antonia Malchik wrote: Antonia Malchik, *A Walking Life* (New York: Da Capo Press, 2019), 25.

173   Adolph's team has also found: Lana B. Karasik, Karen E. Adolph, Catherine S. Tamis-LeMonda, and Alyssa L. Zuckerman, "Carry On: Spontaneous

Object Carrying in 13-Month-Old Crawling and Walking Infants," *Developmental Psychology* 48, no. 2 (2012), 389–397. Carli M. Heiman, Whitney G. Cole, Do Kyeong Lee, and Karen E. Adolph, "Object Interaction and Walking: Integration of Old and New Skills in Infant Development," *Infancy* 24, no. 4 (2019), 547–569.

173   babies aimlessly wander around the room: See Justine E. Hock, Sinclaire M. O'Grady, and Karen E. Adolph, "It's the Journey, Not the Destination: Locomotor Exploration in Infants," *Developmental Science* (2018), e12740.

174   seeing-impaired children: Miriam Norris, Patricia J. Spaulding, and Fern H. Brodie, *Blindness in Children* (Chicago: University of Chicago Press, 1957).

174   "How do you learn to walk?": Karen E. Adolph et al., "How Do You Learn to Walk? Thousands of Steps and Dozens of Falls per Day," *Psychological Science* 23, no. 11 (2012), 1387–1394.

175   The average toddler takes: Adolph, "How Do You Learn to Walk?"

175   Even so, they don't begin to walk: David Sutherland, Richard Olshen, and Edmund Biden, *The Development of Mature Walking* (London: Mac Keith Press, 1988).

176   For example, newborn human babies: Jeremy M. DeSilva, Corey M. Gill, Thomas C. Prang, Miriam A. Bredella, and Zeresenay Alemseged, "A Nearly Complete Foot from Dikika, Ethiopia, and Its Implications for the Ontogeny and Function of *Australopithecus afarensis*," *Science Advances* 4, no. 7 (2018), eaar7723. Craig A. Cunningham and Sue M. Black, "Anticipating Bipedalism: Trabecular Organization in the Newborn Ilium," *Journal of Anatomy* 214, no. 6 (2009), 817–829.

176   daily stresses kids put them through: Experimental work with nonhumans has also revealed ways in which bones respond to the novel stresses of bipedal locomotion. In 1939, a goat was born with no forelegs and hopped on two legs. It died in an accident a year later and was examined by Everhard Johannes Slijper, a comparative anatomist at the University of Utrecht. Slijper's goat had skeletal changes to its spine, pelvis, and lower limbs, thought to be a result of its unusual locomotion. Everhard J. Slijper, "Biologic-Anatomical Investigations on the Bipedal Gait and Upright Posture in Mammals, with Special Reference to a Little Goat, Born Without Forelegs," *Proceedings of the Koninklijke Nederlandse Akademie van Wetenschappen* 45 (1942), 288–295. More recently, a Japanese research team trained a macaque monkey to walk on two legs. Like humans, it developed lumbar lordosis. But while humans develop lordosis because their bones and intervertebral discs become wedge-shaped, the macaque only exhibited changes in the discs. Masato Nakatsukasa, Sugio Hayama, and Holger Preuschoft, "Postcranial Skeleton of a Macaque Trained for Bipedal Standing and Walking and Implications for Functional Adaptation," *Folia Primatologica* 64, no. 1–2 (1995), 1–9. In 2020, Gabrielle Russo of Stony Brook University conducted a controlled experiment in which rats were harnessed and encouraged to walk bipedally. Compared with quadrupedal rats, they developed a more forward-placed foramen magnum, lumbar lordosis, and larger leg joints. Gabrielle A. Russo, D'Arcy Marsh, and Adam D. Foster, "Response of the Axial Skeleton to Bipedal Loading Behaviors in an Experimental Animal Model," *Anatomical Record* 303, no. 1 (2020), 150–166.

176   he wobbled from side to side: It seems, then, that toddlers walk like upright apes with bent hips, bent knees, and a wide stance. Chimpanzees, though, appear to develop their gait in a seemingly opposite manner. Chimpanzees are their most bipedal when they are infants (0.1–5.0 years old). At this young age, they are three times more bipedal than adult chimpanzees, spending 6 percent of their time moving on two legs. Lauren Sarringhaus, Laura Mac-Latchy, and John Mitani, "Locomotor and Postural Development of Wild Chimpanzees," *Journal of Human Evolution* 66 (2014), 29–38.

177   Individuals who are paraplegic: Christine Tardieu, "Ontogeny and Phylogeny of Femoro-Tibial Characters in Humans and Hominid Fossils: Functional Influence and Genetic Determinism," *American Journal of Physical Anthropology* 110 (1999), 365–377.

177   The remarkable thing about the lateral: Yann Glard et al., "Anatomical Study of Femoral Patellar Groove in Fetus," *Journal of Pediatric Orthopaedics* 25, no. 3 (2005), 305–308.

178   The results were always the same: Karen E. Adolph, Sarah E. Berger, and Andrew J. Leo, "Developmental Continuity? Crawling, Cruising, and Walking," *Developmental Science* 14, no. 2 (2011), 306–318. See additional references in Adolph and R. Robinson, "The Road to Walking."

CHAPTER 11: BIRTH AND BIPEDALISM

179   "These hips are mighty hips": Lucille Clifton, "Homage to My Hips," *Two-Headed Woman* (Amherst: University of Massachusetts Press, 1980).

180   These last two activities: Alexander Marshack, "Exploring the Mind of Ice Age Man," *National Geographic* 147 (1975), 85. Francesco d'Errico, "The Oldest Representation of Childbirth," in *An Enquiring Mind: Studies in Honor of Alexander Marshack*, ed. Paul G. Bahn (Oxford and Oakville, CT: American School of Prehistoric Research, 2009), 99–109.

181   We don't know too much: But see Pamela Heidi Douglas, "Female Sociality During the Daytime Birth of a Wild Bonobo at Luikotale, Democratic Republic of Congo," *Primates* 55 (2014), 533–542. Birth assistance has been observed in some monkeys. See Bin Yang, Peng Zhang, Kang Huang, Paul A. Garber, and Bao-Guo Li, "Daytime Birth and Postbirth Behavior of Wild *Rhinopithecus roxellana* in the Qinling Mountains of China," *Primates* 57 (2016), 155–160. Wei Ding, Le Yang, and Wen Xiao, "Daytime Birth and Parturition Assistant Behavior in Wild Black-and-White Snub-Nosed Monkeys (*Rhinopithecus bieti*) Yunnan, China," *Behavioural Processes* 94 (2013), 5–8.

181   The baby proceeds unimpeded: Hirata et al. presented evidence that chimpanzees occasionally deviate from this description of birth. Satoshi Hirata, Koki Fuwa, Keiko Sugama, Kiyo Kusunoki, and Hideko Takeshita, "Mechanism of Birth in Chimpanzees: Humans Are Not Unique Among Primates," *Biology Letters* 7, no. 5 (2011), 286–288. See also James H. Elder and Robert M. Yerkes, "Chimpanzee Births in Captivity: A Typical Case History and Report of Sixteen Births," *Proceedings of the Royal Society of London B* 120 (1936), 409–421.

181   On average, labor lasts fourteen hours: Karen Rosenberg, "The Evolution of Modern Human Childbirth," *Yearbook of Physical Anthropology* 35, no. S15 (1992), 89–124.

181 The solution I, and just: There is variation in birth mechanics. See Dana Walrath, "Rethinking Pelvic Typologies and the Human Birth Mechanism," *Current Anthropology* 44 (2003), 5–31.

181 In 1951, University of Pennsylvania: Wilton M. Krogman, "The Scars of Human Evolution," *Scientific American* 184 (1951), 54–57.

182 We can tell from the shape of Lucy's: Christine Berge, Rosine Orban-Segebarth, and Peter Schmid, "Obstetrical Interpretation of the Australopithecine Pelvic Cavity," *Journal of Human Evolution* 13, no. 7 (1984), 573–584. Robert G. Tague and C. Owen Lovejoy, "The Obstetric Pelvis of A.L. 288-1 (Lucy)," *Journal of Human Evolution* 15 (1986), 237–255. Jeremy M. DeSilva, Natalie M. Laudicina, Karen R. Rosenberg, and Wenda R. Trevathan, "Neonatal Shoulder Width Suggests a Semirotational, Oblique Birth Mechanism in *Australopithecus afarensis*," *Anatomical Record* 300 (2017), 890–899.

183 In most female pelvises: Cara M. Wall-Scheffler, Helen K. Kurki, and Benjamin M. Auerbach, *The Evolutionary Biology of the Pelvis: An Integrative Approach* (Cambridge: Cambridge University Press, 2020).

183 "Navigating the birth canal": In Jennifer Ackerman, "The Downsides of Upright," *National Geographic* 210, no. 1 (2006), 126–145.

183 "Even if my head would go through": Lewis Carroll, *Alice's Adventures in Wonderland* (New York: Macmillan, 1865).

184 For them, as with every human culture: Wenda R. Trevathan, *Human Birth: An Evolutionary Perspective* (New York: Aldine de Gruyter, 1987). Karen R. Rosenberg and Wenda R. Trevathan, "Bipedalism and Human Birth: The Obstetrical Dilemma Revisited," *Evolutionary Anthropology* 4 (1996), 161–168. Karen R. Rosenberg and Wenda R. Trevathan, "The Evolution of Human Birth," *Scientific American* 285 (2001), 72–77. Wenda R. Trevathan, *Ancient Bodies, Modern Lives* (Oxford: Oxford University Press, 2010). Also, midwifery is not just about having an extra set of hands ready to catch the baby. Della Campbell, a professor in the School of Nursing at the University of Delaware, compiled data from six hundred human births. Half of the women giving birth were accompanied by a close female friend or family member; the other half were not. Those who had a female companion, known often as a "doula," shortened their labors by over an hour. This benefited not only the mother but the baby as well. Apgar scores, which measure the health of the newborn, were better in the babies born in the presence of a doula. More recently, University of Toronto professor emeritus Ellen Hodnett reviewed twenty-two studies examining over 15,000 births all over the world. Social support during labor, whether it happens in Iran, Nigeria, Botswana, or the United States, shortened its length and reduced both the need for medication and the chance of an emergency C-section. Our bodies are physiologically adapted to have helpers present at birth, and having these helpers present lowers the chance of something going wrong. See Della Campbell, Marian F. Lake, Michele Falk, and Jeffrey R. Backstrand, "A Randomized Control Trial of Continuous Support in Labor by a Lay Doula," *Gynecologic & Neonatal Nursing* 35, no. 4 (2006), 456–464. Ellen D. Hodnett, Simon Gates, G. Justus Hofmeyr, and Carol Sakala, "Continuous Support for Women During Childbirth," *Cochrane Database of Systematic Reviews* 7 (2013). See also work by University of Minnesota School of Public Health professor Katy Kozhimannil.

184   "Childbirth is beautiful": Angela Garbes, *Like a Mother: A Feminist Journey Through the Science and Culture of Pregnancy* (New York: HarperCollins, 2018), 101.

184   Worldwide, nearly 300,000: "Maternal Mortality," World Health Organization, September 19, 2019, https://www.who.int/news-room/fact-sheets/detail/maternal-mortality.

184   According to a 2019 United Nations: Elizabeth O'Casey, "42nd Session of the UN Human Rights Council. General Debate Item 3," United Nations Human Rights Council, September 9–27, 2019.

184   In countries where the average: Using raw data from Max Roser and Hannah Ritchie, "Maternal Mortality," *Our World in Data,* https://ourworldindata.org/maternal-mortality#. "List of Countries by Age at First Marriage," Wikipedia, https://en.wikipedia.org/wiki/List_of_countries_by_age_at_first_marriage.

185   In the United States, about 700: Donna L. Hoyert and Arialdi M. Miniño, "Maternal Mortality in the United States: Changes in Coding, Publication, and Data Release, 2018," *National Vital Statistics Report* 69, no. 2 (2020), 1–16. GBD 2015 Maternal Mortality Collaborators, "Global, Regional, and National Levels of Maternal Mortality, 1990–2015: A Systematic Analysis for the Global Burden of Disease Study 2015," *The Lancet* 388 (2016), 1775–1812.

185   Institutional racism at many points in the process: See Andreea A. Creanga, Brian T. Bateman, Jill M. Mhyre, Elena Kuklina, Alexander Shilkrut, and William M. Callaghan, "Performance of Racial and Ethnic Minority-Serving Hospitals on Delivery-Related Indicators," *American Journal of Obstetrics and Gynecology* 211, no. 6 (2014), 647. e1-16. Jeanne L. Alhusen, Kelly M. Bower, Elizabeth Epstein, and Phyllis Sharps, "Racial Discrimination and Adverse Birth Outcomes: An Integrative Review," *Journal of Midwifery and Women's Health* 61, no. 6 (2016), 707–720. Christopher W. Kuzawa and Elizabeth Sweet, "Epigenetics and the Embodiment of Race: Developmental Origins of US Racial Disparities in Cardiovascular Health," *American Journal of Human Biology* 21, no. 1 (2009), 2–15.

186   This new approach to anthropology: Sherwood L. Washburn, "The New Physical Anthropology," *Transactions of the New York Academy of Sciences* 13, no. 7 (1951), 298–304.

186   In 1960, Sherry Washburn: Sherwood L. Washburn, "Tools and Human Evolution," *Scientific American* 203 (1960), 62–75.

187   In his influential book *Sapiens*: Yuval Noah Harari, *Sapiens: A Brief History of Humankind* (New York: HarperCollins, 2015), 10.

188   In their 2012 study, Dunsworth: Holly Dunsworth, Anna G. Warrener, Terrence Deacon, Peter T. Ellison, and Herman Pontzer, "Metabolic Hypothesis for Human Altriciality," *Proceedings of the National Academy of Sciences* 109, no. 38 (2012), 15212–15216. Dunsworth called this the EGG hypothesis (Energetics, Growth, Gestation).

188   Human newborns' brains average: Jeremy M. DeSilva and Julie J. Lesnik, "Brain Size at Birth Throughout Human Evolution: A New Method for Estimating Neonatal Brain Size in Hominins," *Journal of Human Evolution* 55 (2008), 1064–1074.

188   It would require only a couple: See Herman T. Epstein, "Possible Metabolic

Constraints on Human Brain Weight at Birth," *American Journal of Physical Anthropology* 39 (1973), 135–136.

189 In 2015, Warrener reported: Anna Warrener, Kristi Lewton, Herman Pontzer, and Daniel Lieberman, "A Wider Pelvis Does Not Increase Locomotor Cost in Humans, with Implications for the Evolution of Childbirth," *PLOS ONE* 10, no. 3 (2015), e0118903.

190 Women from the Hadza: Frank W. Marlowe, "Hunter-Gatherers and Human Evolution," *Evolutionary Anthropology* 14 (2005), 54–67. Charles E. Hilton and Russell D. Greaves, "Seasonality and Sex Differences in Travel Distance and Resource Transport in Venezuelan Foragers," *Current Anthropology* 49, no. 1 (2008), 144–153.

190 In 2007, Katherine Whitcome: Katherine K. Whitcome, Liza J. Shapiro, and Daniel E. Lieberman, "Fetal Load and the Evolution of Lumbar Lordosis in Bipedal Hominins," *Nature* 450 (2007), 1075–1078. In addition to the wedging of the vertebrae, the facets connecting one backbone to the next are also angled more obliquely in women. This is thought to provide stability in a back that is more curved and thus more susceptible to injury.

191 Meanwhile, Wall-Scheffler: Cara Wall-Scheffler, "Energetics, Locomotion, and Female Reproduction: Implications for Human Evolution," *Annual Review of Anthropology* 41 (2012), 71–85. Cara M. Wall-Scheffler and Marcella J. Myers, "The Biomechanical and Energetic Advantage of a Mediolaterally Wide Pelvis in Women," *Anatomical Record* 300, no. 4 (2017), 764–775.

191 Walking while carrying an object: Cara M. Wall-Scheffler, K. Geiger, and Karen L. Steudel-Numbers, "Infant Carrying: The Role of Increased Locomotor Costs in Early Tool Development," *American Journal of Physical Anthropology* 133, no. 2 (2007), 841–846.

192 Wall-Scheffler, Whitcome, and other researchers: Wall-Scheffler and Myers, "The Biomechanical and Energetic Advantage of a Mediolaterally Wide Pelvis in Women." Katherine K. Whitcome, E. Elizabeth Miller, and Jessica L. Burns, "Pelvic Rotation Effect on Human Stride Length: Releasing the Constraint of Obstetric Selection," *Anatomical Record* 300, no. 4 (2017), 752–763. Laura T. Gruss, Richard Gruss, and Daniel Schmid, "Pelvic Breadth and Locomotor Kinematics in Human Evolution," *Anatomical Record* 300, no. 4 (2017), 739–751. See also Yoel Rak, "Lucy's Pelvic Anatomy: Its Role in Bipedal Gait," *Journal of Human Evolution* 20 (1991), 283–290.

192 One idea is that the high mortality: Jonathan C. K. Wells, Jeremy M. DeSilva, and Jay T. Stock, "The Obstetric Dilemma: An Ancient Game of Russian Roulette, or a Variable Dilemma Sensitive to Ecology?" *Yearbook of Physical Anthropology* 149, no. S55 (2012), 40–71.

192 Other researchers suggest that the problem: Christopher B. Ruff, "Climate and Body Shape in Hominid Evolution," *Journal of Human Evolution* 21, no. 2 (1991), 81–105. Laura T. Gruss and Daniel Schmitt, "The Evolution of the Human Pelvis: Changing Adaptations to Bipedalism, Obstetrics, and Thermoregulation," *Philosophical Transactions of the Royal Society B* 370, no. 1663 (2015). See also review in Lia Betti, "Human Variation in Pelvis Shape and the Effects of Climate and Past Population History," *Anatomical Record* 300, no. 4 (2017), 687–697.

193   Another hypothesis involves the anatomical: But see Anna Warrener, Kristin
       Lewton, Herman Pontzer, and Daniel Lieberman, "A Wider Pelvis Does Not
       Increase Locomotor Cost in Humans, with Implications for the Evolution of
       Childbirth," *PLOS ONE* 10, no. 3 (2015), e0118903. They hypothesize that
       the higher incidence of ACL injuries in women results from having less mus-
       cular strength than men. In part, this could be because of differences in how
       the sexes are encouraged (or discouraged) from participating in sports at a
       young age.

193   risk of debilitating anterior cruciate ligament tears: See the relationship be-
       tween valgus knee and risk of ACL injury in Mary Lloyd Ireland, "The Fe-
       male ACL: Why Is It More Prone to Injury?" *Orthopaedic Clinics of North
       America* 33, no. 4 (2002), 637–651.

193   A final hypothesis has been pitched: Wenda Trevathan, "Primate Pelvic
       Anatomy and Implications for Birth," *Philosophical Transactions of the Royal
       Society B* 370, no. 1663 (2015). See also Alik Huseynov et al., "Developmen-
       tal Evidence for Obstetric Adaptation of the Human Female Pelvis," *Pro-
       ceedings of the National Academy of Sciences* 113, no. 19 (2016), 5227–5232.
       Ekaterina Stansfield, Krishna Kumar, Philipp Mitteroecker, and Nicole
       Grunstra, "Biomechanical Trade-Offs in the Pelvic Floor Constrain the Evo-
       lution of the Human Birth Canal," *Proceedings of the National Academy of Sci-
       ences* 118, no. 16 (2021), e2022159118.

193   pelvic prolapse impacts 50 percent: See Donna Mazloomdoost, Catrina C.
       Crisp, Steven D. Kleeman, and Rachel N. Pauls, "Primate Care Providers'
       Experience, Management, and Referral Patterns Regarding Pelvic Floor Dis-
       orders: A National Survey," *International Urogynecology Journal* 29 (2018),
       109–118, and references therein.

194   Now, a new goal is in sight: Kipchoge did run 26.2 miles under two hours in
       2019, but it was not during an official race.

194   The year Peters set his marathon: Marathon records from "Marathon World
       Record Progression," Wikipedia, https://en.wikipedia.org/wiki/Marathon
       _world_record_progression.

195   Poor Richard Ellsworth: See Hailey Middlebrook, "Woman Wins 50K Ultra
       Outright, Trophy Snafu for Male Winner Follows," *Runner's World,* August
       15, 2019, https://www.runnersworld.com/news/a28688233/ellie-pell-wins
       -green-lakes-endurance-run-50k.

196   more resistant to fatigue than men's: See, for example, John Temesi et al.,
       "Are Females More Resistant to Extreme Neuromuscular Fatigue?" *Medicine
       & Science in Sports & Exercise* 47, no. 7 (2015), 1372–1382.

196   Author Rebecca Solnit calls: Rebecca Solnit, *Wanderlust: A History of Walking*
       (New York: Penguin Books, 2000), 43. Genesis: "In sorrow thou shalt bring
       forth children."

197   Holly Dunsworth, the University of Rhode Island anthropologist: See Holly
       Dunsworth, "The Obstetrical Dilemma Unraveled," in *Costly and Cute: Help-
       less Infants and Human Evolution*, ed. Wenda Trevathan and Karen Rosenberg
       (Santa Fe: University of New Mexico Press, published in association with
       School for Advanced Research Press, 2016), 29.

CHAPTER 12: GAIT DIFFERENCES AND WHAT THEY MEAN

199 "High'st queen of state": William Shakespeare, *The Tempest*, www.shakespeare .mit.edu/tempest/full.html.

199 In 1977, Wesleyan University: James E. Cutting and Lynn T. Kozlowski, "Recognizing Friends by Their Walk: Gait Perception Without Familiarity Cues," *Bulletin of the Psychonomic Society* 9 (1977), 353–356.

200 Since then, repeated studies: Sarah V. Stevenage, Mark S. Nixon, and Kate Vince, "Visual Analysis of Gait as a Cue to Identity," *Applied Cognitive Psychology* 13, no. 6 (1999), 513–526. Fani Loula, Sapna Prasad, Kent Harber, and Maggie Shiffrar, "Recognizing People from Their Movement," *Journal of Experimental Psychology: Human Perception and Performance* 31, no. 1 (2005), 210–220. Noa Simhi and Galit Yovel, "The Contribution of the Body and Motion to Whole Person Recognition," *Vision Research* 122 (2016), 12–20.

200 In her 2017 study, for example: Carina A. Hahn and Alice J. O'Toole, "Recognizing Approaching Walkers: Neural Decoding of Person Familiarity in Cortical Areas Responsive to Faces, Bodies, and Biological Motion," *NeuroImage* 146, no. 1 (2017), 859–868.

200 When the walkers were close enough: While not specifically about a walk, Beatrice de Gelder and her colleagues published a study in 2005 in which participants were shown images of facial expressions superimposed on mismatched bodies. A welcoming face was placed on a body with a threatening posture and vice versa. The question was whether our first reaction was to react to the face or to the body posture. The answer, surprisingly to me, was that the participants reacted more often to the body posture than to the facial expression. Hanneke K. M. Meeren, Corné C. R. J. van Heijnsbergen, and Beatrice de Gelder, "Rapid Perceptual Integration of Facial Expression and Emotional Body Language," *Proceedings of the National Academy of Sciences* 102, no. 45 (2005), 16518–16523.

200 Research shows that these inferences: Shaun Halovic and Christian Kroos, "Not All Is Noticed: Kinematic Cues of Emotion-Specific Gait," *Human Movement Science* 57 (2018), 478–488. Claire L. Roether, Lars Omlor, Andrea Christensen, and Martin A. Giese, "Critical Features for the Perception of Emotion from Gait," *Journal of Vision* 9, no. 6 (2009), 1–32. See also a foundational study investigating this question: Joann M. Montepare, Sabra B. Goldstein, and Annmarie Clausen, "The Identification of Emotions from Gait Information," *Journal of Nonverbal Behavior* 11 (1987), 33–42.

200 A 2012 study out of Durham University: John C. Thoresen, Quoc C. Vuong, and Anthony P. Atkinson, "First Impressions: Gait Cues Drive Reliable Trait Judgements," *Cognition* 124, no. 3 (2012), 261–271.

200 In a 2013 study, Angela Book: Angela Book, Kimberly Costello, and Joseph A. Camilleri, "Psychopathy and Victim Selection: The Use of Gait as a Cue to Vulnerability," *Journal of Interpersonal Violence* 28, no. 11 (2013), 2368–2383.

201 As Book pointed out: In her paper, Book cites Ronald M. Holmes and Stephen T. Holmes, *Serial Murder* (New York: Sage, 2009).

202 Costilla-Reyes identified twenty-four: Omar Costilla-Reyes, Ruben Vera-Rodriguez, Patricia Scully, and Krikor B. Ozanyan, "Analysis of Spatio-Temporal Representations for Robust Footstep Recognition with Deep

Residual Neural Networks," *IEEE Transactions on Pattern Analysis and Machine Intelligence* 41, no. 2 (2018), 285–296.

203   One of the first symptoms: Joe Verghese et al., "Abnormality of Gait as a Predictor of Non-Alzheimer's Dementia," *New England Journal of Medicine* 347, no. 22 (2002), 1761–1768. Louis M. Allen, Clive G. Ballard, David J. Burn, and Rose Anne Kenny, "Prevalence and Severity of Gait Disorders in Alzheimer's and Non-Alzheimer's Dementias," *Journal of the American Geriatrics Society* 53, no. 10 (2005), 1681–1687.

203   In 2012, Marios Savvides: Jim Giles, "Cameras Know You by Your Walk," *New Scientist* (September 12, 2012), https://www.newscientist.com/article /mg21528835-600-cameras-know-you-by-your-walk. Joseph Marks, "The Cybersecurity 202: Your Phone Could Soon Recognize You Based on How You Move or Walk," *Washington Post* (February 26, 2019), https://www .washingtonpost.com/news/powerpost/paloma/the-cybersecurity-202/2019 /02/26/the-cybersecurity-202-your-phone-could-soon-recognize-you-based -on-how-you-move-or-walk/5c744b9b1b326b71858c6c39.

204   Jeffrey Hausdorff of Tel Aviv University: Ari Z. Zivotofsky and Jeffrey M. Hausdorff, "The Sensory Feedback Mechanisms Enabling Couples to Walk Synchronously: An Initial Investigation," *Journal of Neuroengineering and Rehabilitation* 4, no. 28 (2007), 1–5. For a more recent study by this team, see Ari Z. Zivotofsky, Hagar Bernad-Elazari, Pnina Grossman, and Jeffrey M. Hausdorff, "The Effects of Dual Tasking on Gait Synchronization During Over-Ground Side-by-Side Walking," *Human Movement Science* 59 (2018), 20–29.

204   A year after Zivotofsky's study: Niek R. van Ulzen, Claudine J. C. Lamoth, Andreas Daffertshofer, Gün R. Semin, and Peter J. Beck, "Characteristics of Instructed and Uninstructed Interpersonal Coordination While Walking Side-by-Side," *Neuroscience Letters* 432, no. 2 (2008), 88–93.

204   In 2018, Claire Chambers: Claire Chambers, Gaiqing Kong, Kunlin Wei, and Konrad Kording, "Pose Estimates from Online Videos Show That Side-by-Side Walkers Synchronize Movement Under Naturalistic Conditions," *PLOS ONE* 14, no. 6 (2019), e0217861.

204   At just eighteen years old, Stephen: Stephen King (writing as Richard Bachman), *The Long Walk* (New York: Signet Books, 1979). I emailed King and asked how—as a college kid—he knew that making the participants walk four miles per hour (mph) would be more horrifying than having them walk three mph. He didn't. He mistakenly thought four mph was the average human walking speed.

205   A cross-cultural study done: Robert V. Levine and Ara Norenzayan, "The Pace of Life in 31 Countries," *Journal of Cross-Cultural Psychology* 30, no. 2 (1999), 178–205. Interestingly, Levine and Norenzayan found a correlation between the average pace and three variables: average temperature, economic vitality, and the general culture of the country (individualistic or collectivist). Cold countries with strong economies and individualist values had fast-walking people.

205   A 2011 study by researchers: Michaela Schimpl et al., "Association Between Walking Speed and Age in Healthy, Free-Living Individuals Using Mobile Accelerometry—A Cross-Cultural Study," *PLOS ONE* 6, no. 8 (2011), e23299.

206 However, a twist occurs when: Janelle Wagnild and Cara M. Wall-Scheffler, "Energetic Consequences of Human Sociality: Walking Speed Choices Among Friendly Dyads," *PLOS ONE* 8, no. 10 (2013), e76576. Cara Wall-Scheffler and Marcella J. Myers, "Reproductive Costs for Everyone: How Female Loads Impact Human Mobility Strategies," *Journal of Human Evolution* 64, no. 5 (2013), 448–456.

207 "Yes, there was a time when": Geoff Nicholson, *The Lost Art of Walking* (New York: Riverhead Books, 2008), 14.

CHAPTER 13: MYOKINES AND THE COST OF IMMOBILITY

209 "I have two doctors": George M. Trevelyan, *Clio, a Muse: And Other Essays Literary and Pedestrian* (London: Longmans, Green, 1913).

209 "Walking is a superfood": Katy Bowman, *Move Your DNA: Restore Your Health Through Natural Movement* (Washington State: Propriometrics Press, 2014).

210 Habiba Chirchir, a biological anthropologist: Habiba Chirchir et al., "Recent Origin of Low Trabecular Bone Density in Modern Humans," *Proceedings of the National Academy of Sciences* 112, no. 2 (2015), 366–371. Chirchir cautioned to me in an email that there is a large temporal gap in her sample and that it still remains unclear precisely when this change to a more gracile skeleton happened.

210 Tim Ryan, a Pennsylvania State University: Timothy M. Ryan and Colin N. Shaw, "Gracility of the Modern *Homo sapiens* Skeleton Is the Result of Decreased Biomechanical Loading," *Proceedings of the National Academy of Sciences* 112, no. 2 (2015), 372–377. Chirchir confirmed these results soon after in her own study. Habiba Chirchir, Christopher B. Ruff, Juho-Antti Junno, and Richard Potts, "Low Trabecular Bone Density in Recent Sedentary Modern Humans," *American Journal of Physical Anthropology* 162, no. 3 (2017), 550–560. What I refer to throughout this section as bone density is technically a bone volume/area fraction.

210 In fact, humans have lost: Daniela Grimm et al., "The Impact of Microgravity on Bone in Humans," *Bone* 87 (2016), 44–56. See also Riley Black (formerly Brian Switek), *Skeleton Keys: The Secret Life of Bone* (New York: Riverhead Books, 2019), 108.

210 The answer, says Steven Moore: Steven C. Moore et al., "Leisure Time Physical Activity of Moderate to Vigorous Intensity and Mortality: A Large Pooled Cohort Analysis," *PLOS ONE* 9, no. 11 (2012), e1001335.

211 Researchers at the University of Cambridge: Ulf Ekelund et al., "Physical Activity and All-Cause Mortality Across Levels of Overall and Abdominal Adiposity in European Men and Women: The European Prospective Investigation into Cancer and Nutrition Study (EPIC)," *American Journal of Clinical Nutrition* 101, no. 3 (2015), 613–621.

211 University of Copenhagen physiologist: Bente Klarlund Pedersen, "Making More Minds Up to Move," *TEDx Copenhagen*, September 18, 2012, https://tedxcopenhagen.dk/talks/making-more-minds-move.

212 One in eight American women: "Breast Cancer Facts & Figures 2019–2020," American Cancer Society (Atlanta: American Cancer Society, Inc., 2019). "Breast Cancer," World Health Organization, https://www.who.int/cancer/detection/breastcancer/en/index1.html.

212   But a daily walk reduces: Janet S. Hildebrand, Susan M. Gapstur, Peter T. Campbell, Mia M. Gaudet, and Alpa V. Patel, "Recreational Physical Activity and Leisure-Time Sitting in Relation to Postmenopausal Breast Cancer Risk," *Cancer Epidemiology and Prevention Biomarkers* 22, no. 10 (2013), 1906–1912.

212   exercise lowers the levels: Kaoutar Ennour-Idrissi, Elizabeth Maunsell, and Caroline Diorio, "Effect of Physical Activity on Sex Hormones in Women: A Systematic Review and Meta-Analysis of Randomized Controlled Trials," *Breast Cancer Research* 17, no. 139 (2015), 1–11.

212   Anne McTiernan's team: Anne McTiernan et al., "Effect of Exercise on Serum Estrogens in Postmenopausal Women," *Cancer Research* 64, no. 8 (2004), 2923–2928.

212   Even if a mutation occurs: Stephanie Whisnant Cash et al., "Recent Physical Activity in Relation to DNA Damage and Repair Using the Comet Assay," *Journal of Physical Activity and Health* 11, no. 4 (2014), 770–778.

212   In a study of nearly 5,000: Crystal N. Holick et al., "Physical Activity and Survival After Diagnosis of Invasive Breast Cancer," *Cancer Epidemiology, Biomarkers & Prevention* 17, no. 2 (2008), 379–386. Holick is now the vice president of research operations at HealthCore, Inc.

213   A follow-up by Saudi Arabian: Interestingly, this was only the case for estrogen-response-positive tumor breast cancer. Estrogen-response-negative showed no impact at all—illustrating that the mechanism by which exercise reduces breast cancer risk is through the estrogens. Ezzeldin M. Ibrahim and Abdelaziz Al-Homaidh, "Physical Activity and Survival After Breast Cancer Diagnosis: Meta-Analysis of Published Studies," *Medical Oncology* 28 (2011), 753–765.

213   Similar reductions in recurrence: Erin L. Richman et al., "Physical Activity After Diagnosis and Risk of Prostate Cancer Progression: Data from the Cancer of the Prostate Strategic Urologic Research Endeavor," *Cancer Research* 71, no. 11 (2011), 3889–3895.

213   In fact, a 2016 study: Steven C. Moore et al., "Leisure-Time Physical Activity and Risk of 26 Types of Cancer in 1.44 Million Adults," *JAMA Internal Medicine* 176, no. 6 (2016), 816–825. A 2020 study of three-quarters of a million people found similar results, with moderate exercise reducing the risk of seven different cancers. The cancers include colon (in men), endometrial, myeloma, breast, liver, kidney, and non-Hodgkins lymphoma (in women). Charles E. Matthews et al., "Amount and Intensity of Leisure-Time Physical Activity and Lower Cancer Risk," *Journal of Clinical Oncology* 38 no. 7 (2020), 686–697.

213   In its various forms: "Heart Disease Facts," Centers for Disease Control and Prevention, December 2, 2019, https://www.cdc.gov/heartdisease/facts.htm.

213   A 2002 study of just under: Mihaela Tanasescu et al., "Exercise Type and Intensity in Relation to Coronary Heart Disease in Men," *Journal of the American Medical Association* 288, no. 16 (2002), 1994–2000.

213   Coronary heart disease is all but: David A. Raichlen et al., "Physical Activity Patterns and Biomarkers of Cardiovascular Disease in Hunter-Gatherers," *American Journal of Human Biology* 29, no. 2 (2017), e22919.

214   the average American, according to Nielsen: "Time Flies: U.S. Adults Now

Spend Nearly Half a Day Interacting with Media," Nielsen, July 31, 2018, https://www.nielsen.com/us/en/insights/article/2018/time-flies-us-adults -now-spend-nearly-half-a-day-interacting-with-media.

214 The total daily energy used: Herman Pontzer et al., "Hunter-Gatherer Energetics and Human Obesity," *PLOS ONE* 7, no. 7 (2012), e40503. Herman Pontzer et al., "Constrained Total Energy Expenditure and Metabolic Adaptation to Physical Activity in Adult Humans," *Current Biology* 26, no. 3 (2016), 410–417.

214 A clue is hidden: Many variables factor into this, such as the weight of the person and the speed of walking. There are a couple of different ways to do this math, though there are assumptions built into each. The first is to adopt the standard, but probably flawed, idea of an average adult human "burning" between 70 and 100 kcal/mile walking approximately 3 miles per hour. Assuming 3,500 kcal/pound, which is also flawed but accepted here for the sake of argument, the answer is 40 miles before a loss of one pound. A better approach is to use the Compendium of Physical Activities, which characterizes walking at a moderate pace as 3 MET units (g/kcal/hr). Doing the math here would result in an answer of 70 miles.

214 The currently accepted hypothesis: See Herman Pontzer, "Energy Constraint as a Novel Mechanism Linking Exercise and Health," *Physiology* 33, no. 6 (2018), 384–393. Herman Pontzer, Brian M. Wood, and Dave A. Raichlen, "Hunter-Gatherers as Models in Public Health," *Obesity Reviews* 19, no. S1 (2018), 24–35. Herman Pontzer, "The Crown Joules: Energetics, Ecology, and Evolution in Humans and Other Primates," *Evolutionary Anthropology* 26, no. 1 (2017), 12–24.

215 But chronically high levels: Roberto Ferrari, "The Role of TNF in Cardiovascular Disease," *Pharmacological Research* 40, no. 2 (1999), 97–105.

215 In 2017, Stoyan Dimitrov: The mechanism is this: Walking increases epinephrine and norepinephrine. These activate receptors called beta-s adrenergic receptors (on immune cells), which then downregulate TNF (proinflammatory cytokines). Stoyan Dimitrov, Elaine Hulteng, and Suzi Hong, "Inflammation and Exercise: Inhibition of Monocytic Intracellular TNF Production by Acute Exercise Via $_2$-Adrenergic Activation," *Brain, Behavior, and Immunity* 61 (2017), 60–68.

215 In the late 1990s, a research team: Kenneth Ostrowski, Thomas Rohde, Sven Asp, Peter Schjerling, and Bente Klarlund Pedersen, "Pro- and Anti-Inflammatory Cytokine Balance in Strenuous Exercise in Humans," *Journal of Physiology* 515, no. 1 (1999), 287–291.

215 To figure out what was going on: Adam Steensberg et al., "Production of Interleukin-6 in Contracting Human Skeletal Muscles Can Account for the Exercise-Induced Increase in Plasma Interleukin-6," *Journal of Physiology* 529, no. 1 (2000), 237–242.

216 In 2003, Pedersen coined a name: Bente Klarlund Pedersen et al., "Searching for the Exercise Factor: Is IL-6 a Candidate?" *Journal of Muscle Research and Cell Motility* 24 (2003), 113–119.

216 Pedersen's team also discovered: Line Pedersen et al., "Voluntary Running Suppresses Tumor Growth Through Epinephrine- and IL-6-Dependent NK Cell Mobilization and Redistribution," *Cell Metabolism* 23, no. 3 (2016),

554–562. See Alejandro Lucia and Manuel Ramírez, "Muscling In on Cancer," *New England Journal of Medicine* 375, no. 9 (2016), 892–894.

216   But that does not require walking: T. Kinoshita et al., "Increase in Interleukin-6 Immediately After Wheelchair Basketball Games in Persons with Spinal Cord Injury: Preliminary Report," *Spinal Cord* 51, no. 6 (2013), 508–510. T. Ogawa et al., "Elevation of Interleukin-6 and Attenuation of Tumor Necrosis Factor–Alpha During Wheelchair Half Marathon in Athletes with Cervical Spinal Cord Injuries," *Spinal Cord* 52 (2014), 601–605. Rizzo quote from Antonia Malchik, *A Walking Life* (New York: Da Capo Press, 2019).

217   On average, Americans take 5,117: David R. Bassett, Holly R. Wyatt, Helen Thompson, John C. Peters, and James O. Hill, "Pedometer-Measured Physical Activity and Health Behaviors in U.S. Adults," *Medicine & Science in Sports & Exercise* 42, no. 10 (2010), 1819–1825.

217   Where does this magical 10,000-step: While this section focuses on the 10,000-step threshold, the practice of counting steps has a much deeper history. According to Dartmouth professor of digital humanities and social engagement Jacqueline Wernimont, the first pedometer goes back to the sixteenth century, and even Napoleon counted his steps per doctor's orders. What has changed through time is the number of steps (currently 10,000) connected with health. See Jacqueline D. Wernimont, *Numbered Lives: Life and Death in Quantum Media* (Cambridge, MA: MIT Press, 2019).

217   In Tokyo that year, Abebe Bikila: Bikila was the first sub-Saharan African Olympic gold medalist in the marathon. He famously won the 1960 gold medal in Rome, running the 26.2 miles barefoot. Since his win in 1964, half of all marathon gold medals have gone to runners from Ethiopia, Kenya, or Uganda. Sadly, Bikila was paralyzed in a car accident in 1969 and died in 1973 at the age of only forty-one.

217   The following year, Hatano: See Catrine Tudor-Locke, Yoshiro Hatano, Robert P. Pangrazi, and Minsoo Kang, "Revisiting 'How Many Steps Are Enough?'" *Medicine & Science in Sports & Exercise* 40, no. 7 (2008), S537–S543.

218   From 2011 to 2015, I-Min Lee: I-Min-Lee et al., "Association of Step Volume and Intensity with All-Cause Mortality in Older Women," *JAMA Internal Medicine* 179, no. 8 (2019), 1105–1112.

218   Keeping it simple, Lee: Carey Goldberg, "10,000 Steps a Day? Study in Older Women Suggests 7,500 Is Just as Good for Living Longer," WBUR, May 29, 2019, https://www.wbur.org/commonhealth/2019/05/29/10000-steps -longevity-older-women-study.

218   Dogs were the first animals: Pontus Skoglund, Erik Ersmark, Eleftheria Palkopoulou, and Love Dalén, "Ancient Wolf Genome Reveals an Early Divergence of Domestic Dog Ancestors and Admixture into High-Latitude Breeds," *Current Biology* 25, no. 11 (2015), 1515–1519. Kari Prassack, Josephine DuBois, Martina Lázni ková-Galetová, Mietje Germonpré, and Peter S. Ungar, "Dental Microwear as a Behavioral Proxy for Distinguishing Between Canids at the Upper Paleolithic (Gravettian) Site of Předmostí, Czech Republic," *Journal of Archaeological Science* 115 (2020), 105092.

219   Even today, dog owners average: Philippa M. Dall et al., "The Influence of Dog Ownership on Objective Measures of Free-Living Physical Activity and

Sedentary Behavior in Community-Dwelling Older Adults: A Longitudinal Case-Controlled Study," *BMC Public Health* 17, no. 1 (2017), 1–9.

219   In addition to preventing some cancers: Hikaru Hori, Atsuko Ikenouchi-Sugita, Reiji Yoshimura, and Jun Nakamura, "Does Subjective Sleep Quality Improve by a Walking Intervention? A Real-World Study in a Japanese Workplace," *BMJ Open* 6, no. 10 (2016), e011055. Emily E. Hill et al., "Exercise and Circulating Cortisol Levels: The Intensity Threshold Effect," *Journal of Endocrinological Investigation* 31, no. 7 (2008), 587–591. Jacob R. Sattelmair, Tobias Kurth, Julie E. Buring, and I-Min Lee, "Physical Activity and Risk of Stroke in Women," *Stroke* 41, no. 6 (2010), 1243–1250. This study showed a dose-dependent effect, which means that the amount and pace of walking mattered.

CHAPTER 14: WHY WALKING HELPS US THINK

221   "Moreover, you must walk": Henry David Thoreau, "Walking," *Atlantic Monthly* (1861).

222   Janet Browne, author of a: Janet Browne, *Charles Darwin: The Power of Place* (Princeton, NJ: Princeton University Press, 2002), 402.

224   the answer comes to you: Columbia University psychologist Christine E. Webb has written about walking as the embodiment of "moving on" among other ways we solve problems. Christine E. Webb, Maya Rossignac-Milon, and E. Tory Higgins, "Stepping Forward Together: Could Walking Facilitate Interpersonal Conflict Resolution?" *American Psychologist* 72, no. 4 (2017), 374–385.

224   The nineteenth-century English poet: Rebecca Solnit wrote of Wordsworth in her book *Wanderlust*, "I always think of him as one of the first to employ his legs as an instrument of philosophy." Rebecca Solnit, *Wanderlust: A History of Walking* (New York: Penguin Books, 2000), 82.

224   French philosopher Jean-Jacques Rousseau: Jean-Jacques Rousseau, *Les Confessions* (1782–1789). Quote from Duncan Minshull, *The Vintage Book of Walking* (London: Vintage, 2000), 10.

224   Nietzsche, who walked with: Friedrich Nietzsche, *Götzen-Dämmerung* (Twilight of the Idols, or, How to Philosophize with a Hammer) (Leipzig: C.G. Naumann, 1889).

224   "The road was so lonely": Charles Dickens, *Uncommercial Traveller*, "Chapter 10: Shy Neighborhoods" (London: All the Year Round, 1860).

225   More recently, Robyn Davidson: Robyn Davidson, *Tracks: A Woman's Solo Trek Across 1700 Miles of Australian Outback* (New York: Vintage, 1995).

225   Historically, however, walking: See Solnit, *Wanderlust*, Chapter 14.

225   Oppezzo designed: Marily Oppezzo and Daniel L. Schwartz, "Give Your Ideas Some Legs: The Positive Effect of Walking on Creative Thinking," *Journal of Experimental Psychology: Learning, Memory, and Cognition* 40, no. 4 (2014), 1142–1152.

226   A few years earlier, Michelle Voss: Michelle W. Voss et al., "Plasticity of Brain Networks in a Randomized Intervention Trial of Exercise Training in Older Adults," *Frontiers in Aging Neuroscience* 2 (2010), 1–17. The stretching exercises of the control group make certain that any brain changes were a result of the cardiovascular changes associated with walking, not with the social stimulation of the group class.

226 In 2004, Jennifer Weuve: Jennifer Weuve et al., "Physical Activity, Including Walking, and Cognitive Function in Older Women," *Journal of the American Medical Association* 292, no. 12 (2004), 1454–1461.

230 In 2011, University of Pittsburgh: Kirk Erickson et al., "Exercise Training Increases Size of Hippocampus and Improves Memory," *Proceedings of the National Academy of Sciences* 108, no. 7 (2011), 3017–3022.

230 In 2018, Sophie Carter: Sophie Carter et al., "Regular Walking Breaks Prevent the Decline in Cerebral Blood Flow Associated with Prolonged Sitting," *Journal of Applied Physiology* 125, no. 3 (2018), 790–798.

231 In 2019, researchers at the Federal: Mychael V. Lourenco et al., "Exercise-Linked FNDC5/Irisin Rescues Synaptic Plasticity and Memory Defects in Alzheimer's Models," *Nature Medicine* 25, no. 1 (2019), 165–175.

231 John Ratey, a clinical psychiatry: See John J. Ratey and Eric Hagerman, *Spark: The Revolutionary New Science of Exercise and the Brain* (New York: Little, Brown Spark, 2013). Kirk Erickson, the lead author of the University of Pittsburgh study, said in an email that they were not able to determine that the circulating BDNF in his study participants was derived directly from muscle since other tissues can produce it as well.

231 "I told myself I wasn't doing": Geoff Nicholson, *The Lost Art of Walking* (New York: Riverhead Books, 2008), 32.

231 One in twelve Americans: Descriptions of depression remind me of one of Zeno's paradoxes. Zeno, a fifth-century BC Italian philosopher, had his audiences imagine walking across a courtyard toward a wall on the other side. First, walk half the distance. Then, half of the remaining distance. Then, half of that. If you continue your journey in this way—splitting each remaining distance by half—you can never reach the wall on the other side. Those halves get infinitesimally small, but there is always another half to go. Imagine the frustration, the exhaustion, the utter hopelessness of always coming up short. According to some accounts, however, when Augustine of Hippo, an Algerian-born Christian priest later canonized as St. Augustine, was presented with Zeno's paradox, he had an answer. *Solvitur ambulando*, he said. "It is solved by walking." His expression has become a rallying cry for pragmatists, not too different from Nike's famous slogan "Just do it."

232 Bratman, then a doctoral student: Gregory N. Bratman, J. Paul Hamilton, Kevin S. Hahn, Gretchen C. Daily, and James J. Gross, "Nature Experience Reduces Rumination and Subgenual Prefrontal Cortex Activation," *Proceedings of the National Academy of Sciences* 112, no. 28 (2015), 8567–8572. Bratman is now an assistant professor in the environmental and forest sciences department at the University of Washington.

233 walk where there are trees: Some hypothesize that phytoncides—airborne molecules released by plants—impact human physiology. One study suggested that phytoncides from trees increased immune function. Qing Li et al., "Effect of Phytoncide from Trees on Human Natural Killer Cell Function," *International Journal of Immunopathology and Pharmacology* 22, no. 4 (2009), 951–959. Phytoncides may also be the mechanism behind the Japanese tradition of *shinrin-yoku*, or "forest bathing," though physiologically how this works remains unclear.

233 In his 1951 short story: Ray Bradbury, "The Pedestrian," *The Reporter* (1951).

CHAPTER 15: OF OSTRICH FEET AND KNEE REPLACEMENTS

235 "Time wounds all heels": From the Marx Brothers film *Go West* (1940), though the phrase has been attributed to many, and Groucho was not the first. See Garson O'Toole, "Time Wounds All Heels," Quote Investigator, September 23, 2014, https://quoteinvestigator.com/2014/09/23/heels/.

235 "I choose to walk": Elizabeth Barrett Browning, *Aurora Leigh* (London: J. Miller, 1856).

235 Dr. Hutan Ashrafian, a lecturer: Hutan Ashrafian, "Leonardo da Vinci's Vitruvian Man: A Renaissance for Inguinal Hernias," *Hernia* 15 (2011), 593–594.

235 more than a quarter of all men: "Inguinal Hernia," Harvard Health Publishing (July 2019), https://www.health.harvard.edu/a_to_z/inguinal-hernia-a-to-z.

235 Inguinal hernias are a direct result: Gilbert McArdle, "Is Inguinal Hernia a Defect in Human Evolution and Would This Insight Improve Concepts for Methods of Surgical Repair?" *Clinical Anatomy* 10, no. 1 (1997), 47–55.

236 This odd route that testes take: See Alice Roberts, *The Incredible Unlikeliness of Being: Evolution and the Making of Us* (New York: Heron Books, 2014).

237 "In the future, robots will be able": When Hurst was a teenager, his dad would take him to Colorado State University to watch the annual walking-machine decathlon where college student–built robots competed in ten tasks. In 2000, Hurst entered his design and won.

237 For the last two decades, his: Jonathan Hurst, "Walking and Running: Bio-Inspired Robotics," TEDx OregonStateU, March 16, 2016, https://www.youtube.com/watch?v=khqi6SiXUzQ. In an email, Hurst wrote, "The fundamental truths of legged locomotion apply to any number of legs: 2, 4, 6, whatever. We have focused on bipedal locomotion, but the similarities between quadrupedal and bipedal locomotion are greater than the differences."

238 These muscles still anchor: See Leslie Klenerman, *Human Anatomy: A Very Short Introduction* (Oxford: Oxford University Press, 2015). See also Arthur Keith, "The Extent to Which the Posterior Segments of the Body Have Been Transmuted and Suppressed in the Evolution of Man and Allied Primates," *Journal of Anatomy and Physiology* 37, no. 1 (1902), 18–40.

240 Dr. Rebecca Ford, an ophthalmologist: Rebecca L. Ford, Alon Barsam, Prabhu Velusami, and Harold Ellis, "Drainage of the Maxillary Sinus: A Comparative Anatomy Study in Humans and Goats," *Journal of Otolaryngology—Head and Neck Surgery* 40, no. 1 (2011), 70–74.

240 Paleoanthropologist Bruce Latimer: Ann Gibbons, "Human Evolution: Gain Came with Pain," *Science*, February 16, 2013, https://www.sciencemag.org/news/2013/02/human-evolution-gain-came-pain.

240 Like springs, they help absorb: Eric R. Castillo and Daniel E. Lieberman, "Shock Attenuation in the Human Lumbar Spine During Walking and Running," *Journal of Experimental Biology* 221, no. 9 (2018), jeb177949.

241 We are the only animals: Bruce Latimer, "The Perils of Being Bipedal," *Annals of Biomedical Engineering* 33, no. 1 (2005), 3–6.

242 With each step, a force equivalent: See Darryl D. D'Lima et al., "Knee Joint Forces: Prediction, Measurement, and Significance," *Proceedings of the Institution of Mechanical Engineers, Part H: Journal of Engineering in Medicine* 226, no. 2 (2012), 95–102.

242   In the United States alone, more: Numbers are expected to reach 1.28 mil-
      lion by 2030. Matthew Sloan and Neil P. Sheth, "Projected Volume of Primary
      and Revision Total Joint Arthroplasty in the United States, 2030–2060,"
      Meeting of the American Academy of Orthopaedic Surgeons, March 6,
      2018.

242   In 1951, the New York Yankees: Roger Kahn, *The Era, 1947–1957* (New York:
      Ticknor & Fields, 1993), 289.

243   Close to 200,000: Matthew Gammons, "Anterior Cruciate Ligament Injury,"
      Medscape, June 16, 2016, https://emedicine.medscape.com/article/89442
      -overview.

243   more common in women than in men: See David E. Gwinn, John H. Wilck-
      ens, Edward R. McDevitt, Glen Ross, and Tzu-Cheng Kao, "The Relative
      Incidence of Anterior Cruciate Ligament Injury in Men and Women at the
      United States Naval Academy," *American Journal of Sports Medicine* 28, no. 1
      (2000), 98–102. Danica N. Giugliano and Jennifer L. Solomon, "ACL Tears
      in Female Athletes," *Physical Medicine and Rehabilitation Clinics of North
      America* 18, no. 3 (2007), 417–438.

244   "I hated it," he told: Christa Larwood, "Van Phillips and the Cheetah Pros-
      thetic Leg: The Next Step in Human Evolution," *OneLife Magazine,* no. 19
      (2010).

245   No living mammals have joined: Good summary in Steve Brusatte, *The Rise
      and Fall of the Dinosaurs: The Untold Story of a Lost World* (New York: William
      Morrow, 2018). See also Pincelli M. Hull et al., "On Impact and Volcanism
      Across the Cretaceous-Paleogene Boundary," *Science* 367, no. 6475 (2020),
      266–272.

245   One early skeletal modification: Qiang Ji et al., "The Earliest Known Euthe-
      rian Mammal," *Nature* 416 (2002), 816–822.

246   A million Americans sprain it every year: Shweta Shah et al., "Incidence
      and Cost of Ankle Sprains in United States Emergency Departments," *Sports
      Health* 8, no. 6 (2016), 547–552.

247   ligament damage to most humans: "Most" humans rather than "all" hu-
      mans because some forest-dwelling humans who climb trees to obtain honey
      have longer muscle fibers and a greater range of motion of the ankle joint.
      See Vivek V. Venkataraman, Thomas S. Kraft, and Nathaniel J. Dominy,
      "Tree Climbing and Human Evolution," *Proceedings of the National Academy
      of Sciences* 110, no. 4 (2013), 1237–1242. Thomas S. Kraft, Vivek V. Ven-
      kataraman, and Nathaniel J. Dominy, "A Natural History of Tree Climbing,"
      Journal of Human Evolution 71 (2014), 105–118.

248   These modifications, the biological: François Jacob, "Evolution and Tinker-
      ing," *Science* 196, no. 4295 (1977), 1161–1166.

249   propelling us into our next steps: See Dominic James Farris, Luke A. Kelly,
      Andrew G. Cresswell, and Glen A. Lichtwark, "The Functional Importance
      of Human Foot Muscles for Bipedal Locomotion," *Proceedings of the National
      Academy of Sciences* 116, no. 5 (2019), 1645–1650.

249   known for their exceptional distance-running abilities: Christopher Mc-
      Dougall, *Born to Run: A Hidden Tribe, Superathletes, and the Greatest Race the
      World Has Never Seen* (New York: Vintage, 2009). See also Daniel E. Lieber-
      man et al., "Running in Tarahumara (Rarámuri) Culture: Persistence Hunt-

ing, Footracing, Dancing, Work, and the Fallacy of the Athletic Savage," *Current Anthropology* 61, no. 3 (2020), 356–379.

250 Lieberman and postdoctoral researchers Nicholas Holowka: Nicholas B. Holowka, Ian J. Wallace, and Daniel E. Lieberman, "Foot Strength and Stiffness Are Related to Footwear Use in a Comparison of Minimally- vs. Conventionally-Shod Populations," *Scientific Reports* 8, no. 3679 (2018), 1–12.

250 Elizabeth Miller of the University of Cincinnati's: Elizabeth E. Miller, Katherine K. Whitcome, Daniel E. Lieberman, Heather L. Norton, and Rachael E. Dyer, "The Effect of Minimal Shoes on Arch Structure and Intrinsic Foot Muscle Strength," *Journal of Sport and Health Science* 3, no. 2 (2014), 74–85.

250 resulting in the stabbing pain: See T. Jeff Chandler and W. Ben Kibler, "A Biomechanical Approach to the Prevention, Treatment, and Rehabilitation of Plantar Fasciitis," *Sports Medicine* 15 (1993), 344–352. Daniel E. Lieberman, *The Story of the Human Body: Evolution, Health, and Disease* (New York: Pantheon, 2013).

250 Harvard University biomechanist Irene Davis said: Stephen J. Dubner, "These Shoes Are Killing Me," *Freakonomics Radio*, July 19, 2017, https://freakonomics .com/podcast/shoes/.

250 High-heeled shoes shorten: Robert Csapo et al., "On Muscle, Tendon, and High Heels," *Journal of Experimental Biology* 213 (2010), 2582–2588.

250 Repeatedly squeezing the end: See Michael J. Coughlin and Caroll P. Jones, "Hallux Valgus: Demographics, Etiology, and Radiographic Assessment," *Foot & Ankle International* 28, no. 7 (2007), 759–779. Ajay Goud, Bharti Khurana, Christopher Chiodo, and Barbara N. Weissman, "Women's Musculoskeletal Foot Conditions Exacerbated by Shoe Wear: An Imaging Perspective," *American Journal of Orthopaedics* 40, no. 4 (2011), 183–191. Lieberman, *The Story of the Human Body.*

251 fractured his right ankle: As Dr. Hecht wrote in a follow-up email: "The injury required surgical repair using metal plates and screws. Unfortunately, he developed painful post-traumatic arthritis which required ankle fusion surgery."

CONCLUSION: THE EMPATHETIC APE

253 "What a frail, easily hurt": D. H. Lawrence, *Lady Chatterley's Lover* (Italy: Tipografia Giuntina, 1928).

253 half a million deaths: "Falls," World Health Organization, January 16, 2018, https://www.who.int/news-room/fact-sheets/detail/falls.

255 We aren't sure what species: The number of species of hominins coexisting at any one time is a contentious topic. Koobi Fora, Kenya, 1.9 million years ago is no different. There are at least two at this time: *Homo* and a robust *Australopithecus* called *Australopithecus* (or *Paranthropus*) *boisei*. But there could be up to four species. Some researchers hypothesize that there are two early *Homo* species they call *Homo habilis* and *Homo rudolfensis*. And a 1.9-million-year-old skull fragment named KNM-ER 2598 is assigned to *Homo erectus*, indicating that this taxon had evolved by this point as well, bringing the total number of species at Koobi Fora, Kenya, 1.9 million years ago to four.

255 break their ankles in childhood: Jeremy M. DeSilva and Amanda Papakyrikos, "A Case of Valgus Ankle in an Early Pleistocene Hominin," *International Journal of Osteoarchaeology* 21, no. 6 (2011), 732–742.

257 A 3.4-million-year-old *Australopithecus afarensis*: Yohannes Haile-Selassie et al., "An Early *Australopithecus afarensis* Postcranium from Woranso-Mille, Ethiopia," *Proceedings of the National Academy of Sciences* 107, no. 27 (2010), 12121–12126.

257 KNM-ER 738 broke her left femur: Richard E. F. Leakey, "Further Evidence of Lower Pleistocene Hominids from East Rudolf, North Kenya," *Nature* 231 (1971), 241–245.

257 fossil is named KNM-ER 1808: For hypervitaminosis A because of liver consumption, see Alan Walker, Michael R. Zimmerman, and Richard E. F. Leakey, "A Possible Case of Hypervitaminosis A in *Homo erectus*," *Nature* 296, no. 5854 (1982), 248–250. Alan Walker and Pat Shipman, *The Wisdom of the Bones: In Search of Human Origins* (New York: Vintage, 1997). An alternative hypothesis posits that KNM-ER 1808 ate too much honey, which also contains high concentrations of vitamin A. See Mark Skinner, "Bee Brood Consumption: An Alternative Explanation for Hypervitaminosis A in KNM-ER 1808 (*Homo erectus*) from Koobi Fora, Kenya," *Journal of Human Evolution* 20, no. 6 (1991), 493–503. For the yaws explanation, see Bruce M. Rothschild, Israel Hershkovitz, and Christine Rothschild, "Origin of Yaws in the Pleistocene," *Nature* 378 (1995), 343–344.

257 It is hard to imagine he: Whether KNM-ER 1808 is an osteological male or female is contentious. Based on what appeared to be a wide sciatic notch of the pelvis and a small brow ridge, Walker et al., *Nature*, 1982, hypothesized that 1808 was an osteological female. Given evidence that the criteria used to sex modern human skeletons may not work as well in earlier hominins, the sheer size of the skeleton, and subsequent findings that *Homo erectus* likely had body size dimorphism, I tend to think that 1808 was an osteological male, thus the "he" in this sentence. I may be wrong, of course.

257 The 1.49-million-year-old Nariokotome: Bruce Latimer and James C. Ohman, "Axial Dysplasia in *Homo erectus*," *Journal of Human Evolution* 40 (2001), A12. Another team suggests that Nariokotome did not have scoliosis, but instead trauma-induced disc herniation. See Regula Schiess, Thomas Boeni, Frank Rühli, and Martin Haeusler, "Revisiting Scoliosis in the KNM-WT 15000 *Homo erectus* Skeleton," *Journal of Human Evolution* 67 (2014), 48–59. Martin Haeusler, Regula Schiess, and Thomas Boeni, "Evidence for Juvenile Disc Herniation in a *Homo erectus* Boy Skeleton," *Spine* 38, no. 3 (2013), E123–E128.

257 A 1.8-million-year-old partial foot: Elizabeth Weiss, "Olduvai Hominin 8 Foot Pathology: A Comparative Study Attempting a Differential Diagnosis," *HOMO: Journal of Comparative Human Biology* 63, no. 1 (2012), 1–11. Randy Susman hypothesizes that the lesions on the OH 8 foot are trauma-induced. See Randall L. Susman, "Brief Communication: Evidence Bearing on the Status of *Homo habilis* at Olduvai Gorge," *American Journal of Physical Anthropology* 137, no. 3 (2008), 356–361.

257 Hominin leg bones: Susman, "Brief Communication."

257 Vertebrae from 2.5-million-year-old: Edward J. Odes et al., "Osteopathology

and Insect Traces in the *Australopithecus africanus* Skeleton StW 431," *South African Journal of Science* 113, no. 1–2 (2017), 1–7.

257 In the same cave deposits: G. R. Fisk and Gabriele Macho, "Evidence of a Healed Compression Fracture in a Plio-Pleistocene Hominid Talus from Sterkfontein, South Africa," *International Journal of Osteoarchaeology* 2, no. 4 (1992), 325–332.

258 Karabo, the *Australopithecus sediba*: Patrick S. Randolph-Quinney et al., "Osteogenic Tumor in *Australopithecus sediba*: Earliest Hominin Evidence for Neoplastic Disease," *South African Journal of Science* 112, no. 7–8 (2016), 1–7.

258 Harvard University primatologist Richard Wrangham: Richard Wrangham, *The Goodness Paradox: The Strange Relationship Between Virtue and Violence in Human Evolution* (New York: Vintage, 2019).

258 Scholars have debated the essence: This is often framed as being aligned with Thomas Hobbes (humans as naturally selfish) or with Jean-Jacques Rousseau (humans as naturally good), though Wrangham argues that Rousseau was not as Rousseauian as many believe. See Wrangham, *The Goodness Paradox*, 5, 18. Robert M. Sapolsky, *Behave: The Biology of Humans at Our Best and Worst* (New York: Penguin Press, 2017). Nicholas A. Christakis, *Blueprint: The Evolutionary Origins of a Good Society* (New York: Little, Brown Spark, 2019). Brian Hare and Vanessa Woods, *Survival of the Friendliest: Understanding Our Origins and Rediscovering Our Common Humanity* (New York: Random House, 2020).

258 but they also bite: Numbers on dog bites and human fatalities from "List of Fatal Dog Attacks in the United States," Wikipedia, https://en.wikipedia.org/wiki/List_of_fatal_dog_attacks_in_the_United_States.

259 That day ended uneventfully: For the chimpanzees, that is. This happened to be my wife's first day in the forest with the chimpanzees and already our group had seen them hunt and eat red colobus monkeys. Soon after, we hid in the buttresses of fig trees as a small herd of forest elephants moved through. The chimpanzees then took us down into a swamp where we were knee-deep in thick muck. That was when we disturbed a nest of "killer" bees. Stuck in the mud, it was impossible to run, and the bees mercilessly stung us over and over again. I swatted at my face and my glasses and Red Sox hat were flung into the forest. My wife grabbed my hand and we pulled ourselves out of the swamp and ran to safety. So perhaps I should not have written that the day ended "uneventfully." My kids love this story and imagine that there is a chimpanzee, deep in the Ugandan rainforest, wearing my glasses and rooting for the baseball team from Boston.

259 beat him to death: Combined data from many chimpanzee sites demonstrate that this is not aberrant behavior caused by the presence of humans. Michael L. Wilson et al., "Lethal Aggression in *Pan* Is Better Explained by Adaptive Strategies Than Human Impacts," *Nature* 513 (2014), 414–417. Sarah Hrdy gives perhaps the best analogy for how humans are more tolerant than chimpanzees. After noting that 1.6 billion humans fly each year, she takes readers on a thought experiment: "What if I were traveling with a planeload of chimpanzees? Any one of us would be lucky to disembark with all ten fingers and toes still attached, with the baby still breathing and unmaimed. Bloody earlobes and other appendages would litter the aisles." Sarah Blaffer Hrdy,

Mothers and Others: The Evolutionary Origins of Mutual Understanding (Cambridge, MA: Belknap Press, 2011), 3.

259   but that does not mean bonobos are pacifists: For meat eating, including primates, see Martin Surbeck and Gottfried Hohmann, "Primate Hunting by Bonobos at LuiKotale, Salonga National Park," *Current Biology* 18, no. 19 (2008), R906–R907. For female coalitions, see Nahoko Tokuyama and Takeshi Furuichi, "Do Friends Help Each Other? Pattern of Female Coalition Formation in Wild Bonobos at Wamba," *Animal Behavior* 119 (2016), 27–35.

259   "The potential for good and evil": Wrangham, *The Goodness Paradox*, 6.

260   the oldest to still preserve DNA: Matthias Meyer et al., "Nuclear DNA Sequences from the Middle Pleistocene Sima de los Huesos Hominins," *Nature* 531 (2016), 504–507.

260   she had a severe case of craniosynostosis: Ana Gracia et al., "Craniosynostosis in the Middle Pleistocene Human Cranium 14 from the Sima de los Huesos, Atapuerca, Spain," *Proceedings of the National Academy of Sciences* 106, no. 16 (2009), 6573–6578.

260   beaten to death with a rock: Nohemi Sala et al., "Lethal Interpersonal Violence in the Middle Pleistocene," *PLOS ONE* 10, no. 5 (2015), e0126589.

260   healed bone around the wound: Christoph P. E. Zollikofer, Marcia S. Ponce de León, Bernard Vandermeersch, and François Lévêque, "Evidence for Interpersonal Violence in the St. Césaire Neanderthal," *Proceedings of the National Academy of Sciences* 99, no. 9 (2002), 6444–6448.

260   girl living in Lazaret Cave: See Marie-Antoinette de Lumley, ed., *Les Restes Humains Fossiles de la Grotte du Lazaret* (Paris: CNRS, 2018).

261   from southern China: Xiu-Jie Wu, Lynne A. Schepartz, Wu Liu, and Erik Trinkaus, "Antemortem Trauma and Survival in the Late Middle Pleistocene Human Cranium from Maba, South China," *Proceedings of the National Academy of Sciences* 108, no. 49 (2011), 19558–19562.

261   instances of traumatic violence: Which raises the question about human warfare. So far, however, we have found nothing in the fossil record to suggest that our hominin ancestors engaged in large-scale conflicts. Warfare may not have arisen until some groups of *Homo sapiens* abandoned the hunter-gatherer life and settled into permanent communities, first to raise cattle and then to farm the land. Well-watered pastureland and fertile soil, it seems, became something worth fighting for. For the latest summary of this research, see Nam C. Kim and Marc Kissel, *Emergent Warfare in Our Evolutionary Past* (New York: Routledge, 2018). The oldest evidence of large-scale human violence—the skeletal remains of an ancient massacre at Nataruk on the shore of Lake Turkana, Kenya—was discovered in 2016 by University of Cambridge paleoanthropologist Marta Mirazón Lahr. There, she unearthed skeletons of ten people who had been bound, stabbed, and beaten to death 10,000 years ago. See Mara Mirazón Lahr et al., "Inter-Group Violence Among Early Holocene Hunter-Gatherers of West Turkana, Kenya," *Nature* 529 (2016), 394–398. The site of Jebel Sahaba, Sudan, is also cited as the earliest evidence for warfare. See Fred Wendorf, *Prehistory of Nubia* (Dallas: Southern Methodist University Press, 1968).

261   humans excel at cooperating: Wrangham differentiates between reactive and

proactive aggression. See Richard Wrangham, "Two Types of Aggression in Human Evolution," *Proceedings of the National Academy of Sciences* 115, no. 2 (2018), 245–253. Wrangham, *The Goodness Paradox*.

261 Too often, we've brushed aside the better angels: For more on this, see Christakis, *Blueprint*; Wrangham, *The Goodness Paradox*; Sapolsky, *Behave*; Hare and Woods, *Survival of the Friendliest*; and Steven Pinker, *The Better Angels of Our Nature: Why Violence Has Declined* (New York: Penguin Group, 2015). For more on the role of cooperation in evolution in general, see Ken Weiss and Anne Buchanan, *The Mermaid's Tale: Four Billion Years of Cooperation in the Making of Living Things* (Cambridge, MA: Harvard University Press, 2009).

261 "Your heart does roughly the same thing": Sapolsky, *Behave*, 44.

262 She was with others—her helpers: This reminds me of a Mr. Rogers quote, "When I was a boy and I would see scary things in the news, my mother would say to me, 'Look for the helpers. You will always find people who are helping.'"

262 Her femur contains a sharp arc: Bone described in Donald C. Johanson et al., "Morphology of the Pliocene Partial Hominid Skeleton (A.L. 288-1) from the Hadar Formation, Ethiopia," *American Journal of Physical Anthropology* 57, no. 4 (1982), 403–451. Possible causes inferred with help from Vincent Memoli, pathologist at Dartmouth Hitchcock Medical Hospital, Lebanon, New Hampshire.

262 Lucy also had back problems: Della Collins Cook, Jane E. Buikstra, C. Jean DeRousseau, and Donald C. Johanson, "Vertebral Pathology in the Afar Australopithecines," *American Journal of Physical Anthropology* 60, no. 1 (1983), 83–101. Scheuermann's disease can be treated in children and does not have to be debilitating. In fact, two recent professional athletes have Scheuermann's: NHL hockey player Milan Lucic and MLB baseball player Hunter Pence.

262 In 2017, I worked with anthropologists: Jeremy M. DeSilva, Natalie M. Laudicina, Karen R. Rosenberg, and Wenda R. Trevathan, "Neonatal Shoulder Width Suggests a Semirotational, Oblique Birth Mechanism in *Australopithecus afarensis*," *Anatomical Record* 300, no. 5 (2017), 890–899.

263 "Midwifery . . . is the 'oldest profession'": Karen Rosenberg and Wenda Trevathan, "Birth, Obstetrics, and Human Evolution," *British Journal of Obstetrics and Gynaecology* 109, no. 11 (2002), 1199–1206.

263 In 2018, Elisa Demuru: Elisa Demuru, Pier Francesco Ferrari, and Elisabetta Palagi, "Is Birth Attendance a Uniquely Human Feature? New Evidence Suggests That Bonobo Females Protect and Support the Parturient," *Evolution and Human Behavior* 39, no. 5 (2018), 502–510. Pamela Heidi Douglas, "Female Sociality During the Daytime Birth of a Wild Bonobo at Luikotale, Democratic Republic of Congo," *Primates* 55 (2014), 533–542. Though Brian Hare, who has studied bonobos at the Lola Ya Bonobo sanctuary, cautioned that the bonobos were excited but that this can lead to bad outcomes. He wrote to me that sometimes "females steal babies and won't give them back," though he also notes that this is not in a wild situation where kinship would be higher.

264 hospitals, wheelchairs, and prosthetics: Robotics professor Jonathan Hurst said in a TED Talk that robots and robotic exoskeletons will eventually make

wheelchairs a thing of the past. Hurst said, "Wheelchairs are going to be an anachronism of history." A 3,000-year-old wood-and-leather toe from Egypt is the oldest known prosthetic. One of the Vedas, the Hindu sacred texts dated to 1100–1700 BC, gives an account of Vi pál, a warrior queen who loses her leg in battle and is fitted with an iron replacement. But, long before the ingenuity required to manufacture a limb replacement, came empathy. From Jacqueline Finch, "The Ancient Origins of Prosthetic Medicine," *The Lancet* 377, no. 9765 (2011), 548–549.

264 empathy starts with the "synchronization of bodies": Frans de Waal, "Monkey See, Monkey Do, Monkey Connect," *Discover* (November 18, 2009). Also see Frans de Waal, *Age of Empathy: Nature's Lessons for a Kinder Society* (New York: Broadway Books, 2010).

264 "In regard to bodily size or strength": Darwin, *The Descent of Man*, 156.

265 Al Capone may have said, "Don't mistake my kindness for weakness": This quote is all over the internet, but it remains unsourced and William J. Helmer, author of *The Wisdom of Al Capone*, calls this "dubious at best" on his website: www.myalcaponemuseum.com.

265 "Don't ever mistake . . . my kindness for weakness": The full quote is: "Don't ever mistake my silence for ignorance, my calmness for acceptance, or my kindness for weakness. Compassion and tolerance are not a sign of weakness, but a sign of strength." Again, however, it is unsourced and may be apocryphal.

265 "We walk on two legs": De Waal, *Age of Empathy*, 159.

266 pulled her to safety: Roger Fouts and Stephen Tukel Mills, *Next of Kin: My Conversations with Chimpanzees* (New York: Avon Books, 1997), 179–180. Gorilla rescue from "20 Years Ago Today: Brookfield Zoo Gorilla Helps Boy Who Fell into Habitat," *Chicago Tribune* (August 16, 2016), https://www.chicagotribune.com/news/ct-gorilla-saves-boy-brookfield-zoo-anniversary-20160815-story.html. Orangutan from Emma Reynolds, "This Orangutan Saw a Man Wading in Snake-Infested Water and Decided to Offer a Helping Hand," CNN, February 7, 2020, https://www.cnn.com/2020/02/07/asia/orangutan-borneo-intl-scli/index.html. A more cynical interpretation is that the orangutan was simply reaching out its hand for food. But since the man did not have any food, that seems unlikely. For information on bonobo behavior, see Vanessa Woods, *Bonobo Handshake* (New York: Gotham, 2010) and sources therein.

267 "If you are a two-legged creature": American Museum of Natural History, "Human Evolution and Why It Matters: A Conversation with Leakey and Johanson," YouTube (May 9, 2011), https://www.youtube.com/watch?v=pBZ8o-lmAsg. Margaret Mead was once asked what the earliest evidence for civilization is. Her answer was, "A healed femur." Ira Byock, *The Best Care Possible: A Physician's Quest to Transform Care Through End of Life* (New York: Avery, 2012).

268 "You're an interesting species": *Contact*, directed by Robert Zemeckis, 1997. In the book *Contact* (New York: Simon & Schuster, 1985), Sagan wrote: "There's a lot in there: feelings, memories, instincts, learned behavior, insights, madness, dreams, loves. Love is very important. You're an interesting mix." The screenplay was written by Michael Goldenberg and James V. Hart.

INDEX

Abell, Paul, 94
acacia, 89–91
Aché Life History (Hill and Hurtado), 169
Aché people, 169–70
Achilles tendon, 247–48, 250
Adolph, Karen, 172–75, 178
Aegyptopithecus, 238
African Genesis (Ardrey), 12–13
Ahouta, Djimdoumalbaye, 61, 62, 64
Aiello, Leslie, 140
Alcibiades, 31
alcohol, 76
Alemseged, Zeray, 102, 110, 294n
Al-Homaidh, Abdelaziz, 213
Ali, Muhammad, 171
alligators and crocodiles, 19–23, 35, 107, 238–39, 278n
Almécija, Sergio, 85
American Association for the Advancement of Science, 68
American Museum of Natural History, 28–29, 95, 120, 266
Ancestor's Tale, The (Dawkins and Wong), 35
Andrewsarchus, 29, 279n
Animal Farm (Orwell), 17
animals
    baby, 164–65
    bipedalism in, *see* bipedalism in animals
    breathing and walking in, 141
    domesticated, 218–19
    pregnancy in, 190
    running in, 6
    tool use in, 8, 103
    upright posture in, 27, 30, 33–34
    see also specific types of animals
ankles, 70, 80, 118, 120, 121, 245–48, 295n
    healed fractures in fossils, 256–58
anterior talofibular ligament, 246–48
anxiety and depression, 231–33, 316n
apes, 8–9, 11, 19, 26–27, 30, 32, 38,
    73–77, 110, 122, 142, 176, 236,
    238, 243, 248, 274n, 304n
    aquatic-ape hypothesis, 34–36, 281n
    bipedalism in, xvi, 32–34, 43–44
    birth and parenting in, 164–65, 181,
        182, 293n
    bonobos, *see* bonobos
    chimpanzees, *see* chimpanzees
    common ancestor of monkeys and, 238
    Danuvius guggenmosi, 79–86
    empathy in, 266, 267
    in Europe, 75, 76, 289n
    fossils of, 75, 78–86
    gibbons, 26, 27, 38, 81–82, 84, 238
    gorillas, *see* gorillas
    orangutans, 26, 27, 38, 75, 76, 84,
        187–88, 238, 293n, 324n
    Rudapithecus hungaricus, 82, 86
    use of word, 275n
aquatic-ape hypothesis, 34–36, 281n
archosaurs, 20–23
Ardipithecus, 100, 109, 129, 130, 287n
Ardipithecus kadabba, 61, 68, 71, 128
Ardipithecus ramidus, 67–71, 73, 74, 86,
    127, 288n
"Ardi," 69–70, 74, 127, 129, 288n
Ardrey, Robert, 12–13
argon, 55, 56
arthritis, 241, 257
Asfaw, Berhane, 67, 70
Ashrafian, Hutan, 235
Atapuerca Mountains, 260
Attenborough, David, 34
Augustine, Saint, 316n
Aurora Leigh (Browning), 235
Australia, 27–29
Australopithecus, 10–11, 13, 54, 58, 66,
    67, 86, 102, 103, 119–20, 128,
    130, 133, 138, 142, 153, 154,
    157, 191, 210, 253, 258
    brains of, 109–12, 142–43
    childbirth and, 262–63
    feet of, 113–14, 117–30, 137

Australopithecus (*continued*)
  parenting and, 104–6, 293n
  predators and, 107–8, 111
  tool use in, 102–4
Australopithecus afarensis, 119, 120,
    126, 129, 153, 257
  Dikika Child, 102, 110–11, 142, 294n
  First Family, 52, 54
  Laetoli footprints, xii–xiii, 93–100, 101,
    104, 119, 130, 204, 262, 292n
  Lucy, 21, 49–58, 68–71, 80, 95,
    100–102, 104–6, 109, 118–20, 122,
    125–29, 137, 153, 182, 254, 255, 262,
    288n
Australopithecus africanus, 10, 13, 119,
    120
  Taung child, 9–12, 14, 52, 119, 275n
Australopithecus anamensis, 104, 109,
    119, 128
Australopithecus deyiremeda, 128, 130
Australopithecus garhi, 128
Australopithecus Prometheus, 12, 13
Australopithecus robustus, 119, 120
Australopithecus sediba, 117–26, 132,
    177, 258
Awe, Rokus Due, 152

babies, *see* children and babies
baboons, 32, 34, 107, 108
back, 240–41
  see also spine and vertebrae
Bamford, Marion, 125
Bannister, Roger, 194
bears, xv–xvii, 30, 97–98, 164
beer, 76
Beethoven, Ludwig van, 224
Behave (Sapolsky), 261–62
Behrensmeyer, Kay, 92
Bennett, Kalin, 171–72
Berger, Lee, 113–19, 124, 125, 155–57,
    276n, 294n
Berger, Matthew, 115–16, 126, 258
Bikila, Abebe, 217, 314n
bipedalism
  brain size and, 8, 48, 54, 73, 109–12
  breathing and, 141–43
  as controlled fall, 3, 4, 6
  energy used in, 42–43, 283n
  health effects of, *see* health effects of
    bipedalism
  speed and, 26, 30, 253
bipedalism in animals, xv–xvii, 5, 30,
    32–33
  ape species, xvi, 32–34, 43–44
  bears, xv–xvii, 30, 97–98
  birds, 18–19, 25–26
  chimpanzees, 32–34, 304n
  dinosaurs, 23–26
  gibbons, 81
  kangaroos, 28–30
  reptiles, 17–18, 23
  size and, 29, 30
  tails and, 26–27
bipedalism in humans, evolution of,
    3–15, 30, 280n, 290n
  childbirth and, *see* childbirth
  different types of walking in, 130
  empathy and, 264–68
  explanations for, 31–46
  females and, 45–46
  Jolly's paradox and, 282n
  knuckle-walking in, 43–44, 69–70,
    72–74, 81, 82, 84–86, 186, 290n
  food and, 43–46, 103, 110
  language and, 3, 141–43
  tools and, 7, 8, 12–14, 36, 44, 101–4,
    111–12
  in wooded areas, 69, 81, 84, 85, 125
  see also fossils; *specific genus and proper
    names of fossils*
Bird, Sue, 243
birds, 23, 25, 237, 298n
  bipedalism in, 18–19, 25–26
  crocodiles and, 19–20, 278n
  emus, 28, 245
  flamingos, 246
  ostriches, 25, 245
  young, 164
Bock, Paul, 146
body hair, 73, 105, 108
Böhme, Madelaine, 77–83
Boisvert, Dick, 146–47
Bolt, Usain, 6, 35
bones, fossil, *see* fossils
bones, living, 175–76
  density of, 209–10
  immobility and, 209–10
  stresses on, 176, 303n
bonobos, 11, 26, 32, 38, 40, 46, 58,
    73, 75, 76, 187, 258–59, 263,
    266, 324n
Book, Angela, 200–201
Boule, Marcellin, 48–49, 54, 284n
Bowman, Katy, 209
Bradbury, Ray, 233–34
Brady, Tom, 243

brain, 3, 54, 228–29, 232
of *Australopithecus*, 109–12, 142–43
  connectivity in, 226
  energy used by, 109, 136, 140
  foramen magnum and, 51–52, 63
  hippocampus in, 228–29, 231
  language and, 142
  and recognizing people from their walks, 200
  size of, 8, 47, 48, 54, 73, 109–12, 130, 132, 136, 139–40, 144, 152, 153, 187, 188, 298n
  of Taung child fossil, 9–10
Brain, Charles Kimberlin "Bob," 13
brain-derived neurotrophic factor (BDNF), 231, 316n
Bratman, Greg, 232
breathing, 141–43
Brodmann, Korbinian, 232
Brooks, Alison, 148
Broom, Robert, 11, 54, 66, 120, 287n
Browning, Elizabeth Barrett, 235
Brunet, Michel, 61–65
Bryant, Kobe, 246
Bundy, Ted, 201
Burtele, 127–30

Cameron, Kirk, 19, 38
cancer, 217, 219, 312n
  breast, 211–13, 216, 312n
  myokines and, 216
  prostate, 213
Capone, Al, 265, 324n
carbon, 40–41, 55–56
Carbonell, Eudald, 139
cardiovascular disease, 213, 215, 217, 219
Carnufex carolinensis ("Carolina Butcher"), 20–23, 78, 239
Carter, Sophie, 230
Carvalho, Susana, 44
Cassie, 236–37
Centers for Disease Control and Prevention, U.S., 6, 167
Chambers, Claire, 204
cheetahs, 6, 51, 244, 274n
Chellappa, Rama, 202–3
childbirth, 164–65, 179–97, 253
  in Australopithecus, 262–63
  baby's position in, 183
  infant deaths in, 184
  ischial spines and, 183, 188, 193–94
  maternal deaths in, 184–85, 192–93

  midwives and other helpers in, 179, 180, 262–64, 305n, 306n
  obstetrical dilemma in, 185–87, 189, 191, 193, 194, 197
  pelvis and, 181–84, 188–89, 191, 193, 262, 263
  and women's ability to walk, 188–92, 196–97
children and babies
  bones of, 176
  crawling of, 172, 173, 178
  development of, 167–68
  first steps of, 174, 301n
  IQ tests and, 171
  learning to walk, 163–78
  mothers' carrying of, 191
  newborn, 53, 106, 165, 176
  number of steps taken per hour, 175
  parenting of, 104–6, 293n
  SIDs deaths in, 168
  spine of, 52
  step reflex in, 165–66, 301n
chimpanzees, 5, 11, 26, 38, 44, 46, 51, 63, 75, 76, 82, 84, 85, 108, 119, 142, 176, 177, 186, 201, 210, 238, 248, 259, 261, 263, 290n, 304n, 321n
  Achilles tendon in, 247–48
  baby, 105, 106
  bipedalism in, 32–34, 304n
  birth in, 263
  bones of, 175–76
  brain size of, 109–11, 188
  Darwin on, 264–65
  empathy in, 266
  energy used in movement, 42–43, 109–10
  gestation in, 187
  humans' common ancestor and split from, 3, 19, 38–40, 58, 65, 73, 74, 76, 82–84, 263
  in March of Progress image, 72–73, 95
  Ngogo community of, 247
  pelvis of, 52–53
  tools used by, 8, 103
  violence in, 258–59, 321n
Chirchir, Habiba, 210
Clancy, Kate, 168
Clarke, Arthur C., 7
coccyx, 238
Cofran, Zach, 113–14
cognitive decline, 226–27, 229–31
common ancestors, 19–20

common ancestors (*continued*)
  of apes and monkeys, 238
  of birds and crocodiles, 19–20, 278n
  of humans and chimpanzees, 3, 19,
    38–40, 58, 65, 73, 74, 76, 82–84, 263
Contact (Sagan), 267–68, 325n
Cope, Jason, 34
coprolite, 125
coronary heart disease, 213, 215, 217,
  219
Costilla-Reyes, Omar, 202–3
Cradle of Humankind, 11–13, 154, 156
  Gladysvale, 114–16
creativity, 225–26
Cressman, Luther, 150–51
crocodiles and alligators, 19–23, 35,
  107, 238–39, 278n
Cro-Magnon, 72
Cutting, James, 199

Dalai Lama, 265
Danuvius guggenmosi, 79–86
Dart, Raymond, 9–14, 45, 52, 54, 119,
  275n, 276n
Dartmouth College, 227
Darwin, Charles, 8–9, 12, 33, 45, 104,
  109, 139, 203, 221–22, 264–65,
  275n
  The Descent of Man, 8, 38, 47, 101, 275n
  On the Origin of Species, 7, 8
  Sandwalk of, 222–24
Dauwalter, Courtney, 196
Davidson, Robyn, 225
Davis, Irene, 250
Davis, Jenny, 168
Dawkins, Richard, 35–36
Dechant, Dorothy (Doty), 92
dementia, 226–27, 229–30
Demuru, Elisa, 263
Denisovans, 144, 149, 154, 158, 159
depression and anxiety, 231–33, 316n
Descent of Man, The (Darwin), 8, 38,
  47, 101, 275n
de Waal, Frans, 264, 265
diabetes, 217, 219
Dicerorhinus etruscus, 131–32
Dickens, Charles, 224
Dikika Child, 102, 110–11, 142, 294n
DiMaggio, Joe, 242
Dimitrov, Stoyan, 215, 313n
dinosaurs, xii, 15, 18, 20, 21, 23–26,
  71–72, 245
  T. rex, 20, 23–24, 26, 245, 278n, 279n

Diogenes the Cynic, 3
Dirks, Paul, 115
Disotell, Todd, 38–40, 58, 83–84
Ditsong Natural History Museum, 13,
  277n
Dmanisi, 131–33, 138, 296n
DNA, 19, 38, 39, 58, 82, 108, 143,
  144, 149, 212
dogs, xvi, 10, 11, 238, 258, 266
  domestication of, 218–19
  upright walking by, xvi–xvii
Dominici, Nadia, 166
Douglas, Pamela Heidi, 263
Dove, Heather, 169
Dubois, Eugène, 47–49, 54, 132, 284n
Dunsworth, Holly, 187–88, 197

elephants, 164, 258, 266
Elliott, Marina, 157
Ellsworth, Richard, 195–96
Emerson, Ralph Waldo, 224
empathy, 261–68
emus, 28, 245
energy, 140
  in bipedalism, 42–43, 283n
  brain and, 109, 136, 140
  carrying and, 191
  in children and teens, 136
  daily allowance of, 214
  hip width and, 189
  movement and, 42–43, 109–10
  walking in groups and, 191–92, 205–6
  walking speed and, 205–6
environmental changes, 40–42, 236,
  297n
  ice ages and glaciers, 139, 143–44, 146,
  180
ethanol, 76
Ethiopia, 135
  Middle Awash project in, 70
  National Museum in, 49–50, 128
Eudibamus cursoris, 18
evolution, 19, 65–66, 181–82
  common ancestors in, *see* common
    ancestors
  convergent, 15
  March of Progress image of, 72–73,
    85–86, 95, 288n
  natural selection in, 7–8, 15, 45, 111,
    139, 190, 222, 248
  see also bipedalism in humans, evolution
    of

Fabian, Kallisti, 98
falling, 6, 253
  in bipedal walking, 3, 4, 6
Fannin, Luke, 97
feet, 53–54, 57, 70–71, 80, 159, 176,
    244–46, 295n
  ankles, see ankles
  of apes, 248
  arches of, 137, 138, 248–50
  of Australopithecus, 113–14, 117–30, 137
  barefoot walking and calluses, 91, 291n
  of Homo erectus, 137
  hyperpronation and, 122–25
  plantar aponeurosis in, 249, 250
  shoes and, 150–52, 249–51
  surgery on, 251–52
  toes, 48, 53–54, 61, 69–71, 74, 75, 85,
    86, 102, 151, 248, 249
  see also footprints
femur, 53, 60, 80, 136–37, 176, 177,
    242, 243
fetus, 166, 188
Feuerriegel, Elen, 157
fibula, 246
fire, 140–41, 144, 298n
fish, 236
flamingos, 246
Flores, 152–54
food, 43–46, 103, 108, 110, 137–38,
    140
  cooking of, 140, 298n
footprints
  in Crete, 291n
  differences among people, 202
  Homo erectus, 137
  Laetoli, xii–xiii, 93–100, 101, 104, 119,
    130, 204, 262, 292n
foramen magnum, 51–52, 63
Ford, Rebecca, 240
Fort Rock Cave, 150–51
fossils, xi–xiii, 3, 8–9, 11–14, 19, 22,
    29, 40, 58–59, 62, 65–66, 80, 82,
    86, 114, 228, 236
  of apes, 75, 78–86
  of archosaurs, 20
  at Atapuerca Mountains, 260
  with bite impressions, 14, 107, 253
  at Burtele, 127–30
  coprolite, 125
  dating of, 40, 55–56
  at Dmanisi, 131–33, 138
  at Gladysvale, 114–16

at Hammerschmiede, 77–81, 83
at Laetoli, see Laetoli
at Liang Bua Cave, 152–54
at Malapa Cave, 115–17, 121–22,
    124–26
missing link concept and, 47, 48
at Olorgesailie, 148–49
of predators, 107
at Rising Star Cave, 156–58
scientific community's access to, 60–61,
    64, 80, 124
at Tianyuan Cave, 151
at Woranso-Mille, 126–28, 257
see also specific genus and proper names of
    fossils
Frazier, Joe, 217
From Lucy to Language (Johanson), 21
fruit, 76, 84, 86, 108, 290n
Fuss, Jochen, 78

Gabunia, Leo, 131–32
Galápagos (Vonnegut), 219
Garbes, Angela, 184
genes, 176, 178
see also DNA
genus and species, 10–11, 276n
Gesell, Arnold, 167–68, 301n
gibbons, 26, 27, 38, 81–82, 84, 238
Gigantopithecus, 30
giraffes, 239
glaciers, see ice ages and glaciers
Gladysvale, 114–16
goats, xvi, 34, 240, 303n
Goodall, Jane, 8, 45, 103
Goodness Paradox, The (Wrangham),
    258, 259
gorillas, 11, 26, 38, 63–65, 73, 75, 76,
    82, 84, 238, 273n
  Ambam, 33
  Darwin on, 264–65
  empathy in, 266
  gestation in, 187
  Louis, xvi, 43–44, 95
Gupta, Sanjay, 266
Gurtov, Alia, 157
Gurtu, Josephat, 91, 99
gut, 140

Haddock, Dorris, 225
Hadza, 190, 206, 213, 214, 217
Haile-Selassie, Yohannes, 61, 109
    126–30, 257

Hammerschmiede, 77–81, 83
Han, Carina, 200
hands, 85, 290n
Harari, Yuval Noah, 187
Harcourt-Smith, Will, 120
Hardy, Alister, 34
Harmand, Sonia, 101–3
Harvati, Katerina, 149
Hatano, Yoshiro, 217–18
Hausdorff, Jeffrey, 204
Hay, Dick, 93
Hayes, Bob, 217
health benefits of walking, 207,
    209–19
  cancer and, 211–13, 216, 217, 219, 312n
  cardiovascular disease and, 213, 215,
    217, 219
  life span and, 210–11
  and number of steps per day, 217, 314n
  weight loss and, 214, 313n
health effects of bipedalism, 235–52,
    253–54
  on feet and ankles, 244–52
  inguinal hernias, 235–36
  on knees, 242–44
  on sinus cavities, 239–40
  on spine, 240–41
heart disease, 213, 215, 217, 219
Hecht, Paul, 251–52, 319n
Henry, Amanda, 125
Herron, Camille, 196
Hewes, Gordon, 44–45
Hill, Andrew, 92–93, 291n
Hill, Kim, 169
Hillaby, John, xv
hippocampus, 228–29, 231
hips, see pelvis and hips
Hobbes, Thomas, 321n
"Hobbit, the" (Homo floresiensis), 153,
    154, 158–59
Holick, Crystal, 212
Holowka, Nicholas, 250
Holt, Ken, 122–23
hominids, 11
hominin ancestors, xii, 33–38, 40, 43,
    45–46, 47–48, 54, 58–60, 65–66,
    73–74, 83–84, 130, 186
  birth in, 184
  body hair of, 105, 108
  coexisting species of, 255, 319n
  migration of, 131–33, 136, 138–39, 143,
    148–50, 297n
  small-bodied, 152–54, 158

  see also fossils; specific genus and proper
    names of fossils
hominoids, 274n
Homo, 10, 117, 130, 133, 138, 143–44,
    153, 154
Homo erectus, 47–49, 59, 132, 136–38,
    143, 144, 153, 154, 157, 257,
    284n
  fire and, 140–41, 144
  footprints of, 137
  migration of, 138–39, 143
  Nariokotome Boy, 133–38, 257, 296n
Homo floresiensis ("the Hobbit"), 153,
    154, 158–59
Homo habilis, 94, 101, 102, 157
Homo heidelbergensis, 143–44
Homo luzonensis, 154, 159
Homo naledi, 158, 159
Homo neanderthalensis, see Neandertals
Homo sapiens, 27, 48, 144, 148–49,
    152, 154, 193, 210, 284n
Hunter, Lindsay, 157
Hunter, Rick, 156
Hurst, Jonathan, 237, 317n, 324n
Hurtado, A. Magdalena, 169
Hutchinson, Fred, 213
hyoid bone, 142, 143

Ibrahim, Ezzeldin, 213
ice ages and glaciers, 139, 143–44,
    146, 180
Ig Nobel Committee, 190
immobility, 207, 209–19
immune system, 215, 219
inflammation, 214–15
inguinal hernia, 235–36
interleukin-6, 215–16
IQ tests, 171
irisin, 230–31
ischial spines, 183, 188, 193–94
isotopes, 40–41, 55–56

Janis, Christine, 29
Jobs, Steve, 224
Johanson, Don, 21, 49, 52, 53, 57, 101,
    266, 284n
Johns, Helen, 34
Jones, Peter, 93

kangaroos, 28–30
Kant, Immanuel, 224
Kaplan, Hillard, 169
Karabo, 116, 125, 126, 258

Kayranto, Kampiro, 129
Keats, John, 89
Kenya, 59, 135, 170
    Nairobi, 133–34
    Olorgesailie site, 148–49
Kenya National Museum, 254
Kenyanthropus platyops, 101
Kerouac, Jack, 131
Keyes, Leroy, 171
Kibii, Job, 115
Kilham, Ben, 97
Kimbel, Bill, 124, 125
Kimeu, Kamoya, 135, 136
King, Stephen, 204–5, 310n
Kingdon, Jonathan, 31, 69
Kipchoge, Eliud, 194
Kirshner, Lucy, xii
Kitchel, Nathaniel, 145–47
knees, 53, 80, 121, 122, 176–77
    dislocation of, 123, 177
    hyperpronation and, 122–23
    injuries to, 242–44
KNM-ER 1808 fossil, 257, 320n
KNM-ER 2596 fossil, 254–57
knuckle-walking, 43–44, 69–70,
    72–74, 81, 82, 84–86, 186, 290n
Kosgei, Brigid, 195
Kozlowski, Lynn, 199
Krogman, Wilton, 181–82
Kubrick, Stanley, 7, 13, 43
Kuzawa, Chris, 136

Lady Chatterley's Lover (Lawrence), 253
Laetoli, 89–92
    footprints at, xii–xiii, 93–100, 101, 104,
    119, 130, 204, 262, 292n
Laimbeer, Bill, 17
Lamarck, Jean-Baptiste, 33, 280n
Land of the Lost, 17
language, 3, 141–43
Lao Tzu, 163
Latimer, Bruce, 113, 129–30, 240
Latynina, Larisa, 217
Laudicina, Natalie, 262
Lawrence, D. H., 253
Leakey, Louis, 91, 100–101, 103, 131,
    134, 276n
Leakey, Mary, 90–94, 97, 98, 100–102,
    131, 134
Leakey, Meave, 58, 101
Leakey, Philip, 92, 93
Leakey, Richard, 135, 257, 266–67
Lechner, Thomas, 79

Lee, I-Min, 218
legs
    femur in, 53, 60, 80, 136–37, 176, 177,
    242, 243
    length of, 137, 138, 141, 144, 154
lemurs, 32
Leonardo da Vinci, 235
Le Page, Sally, 124
Levine, Robert V., 205, 310n
Liang Bua Cave, 152–54
Lieberman, Daniel, 190, 250
life span, 210–11
Like a Mother (Garbes), 184
Long Walk, The (King), 204–5, 310n
Lord of the Rings (Tolkien), 145
Lost Art of Walking The (Nicholson), 231
Lovejoy, Owen, 45, 71, 73–74, 282n
Lowly Origin (Kingdon), 69
Lucy, 21, 49–58, 68–71, 80, 95,
    100–102, 104–6, 109, 118–20,
    122, 125–29, 137, 153, 182, 254,
    255, 262, 288n
Lucy (Johanson and Edey), 57
Luzon, 154
Lyell, Charles, 92

Maasai, 89–91, 99–100
MacLatchy, Laura, xii, xiii, 289n
macrophages, 215
Makanku, Justin, 126
Malapa Cave, 115–17, 121–22, 124–26
Malchik, Antonia, 173
Maley, Blaine, 97
Manthi, Fredrick Kyalo, 134
Mantle, Mickey, 242–43
March of Progress, 72–73, 85–86, 95,
    288n
Marx, Groucho, 5, 235
Matalo, Simon, 93
Mays, Willie, 171, 242
Mbuika, Ndibo, 94
McNutt, Ellie, 70, 97–98
McTiernan, Anne, 212
Mead, Margaret, 150
Megatherium, 30
Melillo, Stephanie, 126–29
memory and cognitive decline,
    226–27, 229–31
Mermaids: The Body Found, 34
Middle Awash project, 70
Miller, Elizabeth, 250
Miller, Kate, 99
Millet, Jean-François, 163

Miocene, 77, 78
Miotragocerus, 78
missing link, 47, 48
Mitani, John, 247
monkeys, 27, 32, 45, 73–74, 84, 141,
    279n, 304n
    baboons, 32, 34, 107, 108
    common ancestor of apes and, 238
Moore, Steven, 210–11
Morgan, Alex, 243
Morgan, Elaine, 34
Morris, Hannah, 157
Morton, Dudley J., 81–82
Move Your DNA (Bowman), 209
Muir, John, 224
Museum of Science (Boston), 238–39
Musiba, Charles, 90
myokines, 216–17, 230–31

Nairobi, 133–34
Napier, John, 4
Nariokotome Boy, 133–38, 257, 296n
National Museum of Ethiopia, 49–50,
    128
Nature Research Center, 20–21
Neandertals, 8, 10, 48–49, 144, 149,
    154, 158, 159, 189, 210, 260,
    275n
New England, 145–47, 224
New Physical Anthropology, The
    (Washburn), 186
New York Times, 57
Nicholson, Geoff, 231
Nietzsche, Friedrich, 224
Norenzayan, Ara, 205, 310n

Obama, Barack, 57
octopuses, xvii
Olduvai Gorge, 90, 91, 94, 100–102,
    131
Olorgesailie, 148–49
Olympic Games, 26, 35, 195, 217,
    244, 246
On the Origin of Species (Darwin), 7, 8
On the Road (Kerouac), 131
Oppezzo, Marily, 225–26, 232, 233
orangutans, 26, 27, 38, 75, 76, 84,
    187–88, 238, 293n, 324n
Orrorin tugenensis, 60–61, 63, 68, 71,
    85, 86, 100, 286n
orthopedic surgery, 251–52
Orwell, George, 17
osteoarthritis, 241

ostriches, 25, 245
oxygen, 40, 41

Pääbo, Svante, 149
pair-bonding, 45
paleontology
    anatomy and, 227–28
    paleoanthropology, xi–xiii, 59, 60,
        63–65, 68, 121
Paris, Jasmin, 196
Peabody Museum of Natural History,
    72
Pedersen, Bente Klarlund, 211, 215–16
"Pedestrian, The" (Bradbury), 233–34
"peekaboo" hypothesis, 33, 280n
Peiper, Albrecht, 165
Peixotto, Becca, 157
Pell, Ellie, 196
pelvic organ prolapse, 193–94
pelvis and hips, 52–53, 70, 82, 98,
    120, 121, 176, 243, 253, 285n
    childbirth and, 181–84, 188–89, 191,
        193, 262, 263
    child carrying and, 191
    energy efficiency and, 189
    tailbone and, 238
Peters, Jim, 194, 195
Phelps, Michael, 35, 217
Phillips, Van, 244
Pickering, Robyn, 117
Pickford, Martin, 59–60, 63–64, 286n
Piercy, Violet, 194
Piontelli, Alessandra, 166
Pistorius, Oscar, 244, 245
Pithecanthropus erectus, 47–48, 284n
plantar aponeurosis, 249, 250
plantar fasciitis, 249
Plaquette 59, 180
Plato, 3–4
Pontzer, Herman, 137, 189, 214
Postosuchus, 20
potassium, 55–56
Potts, Rick, 148
Potze, Stephany, 13–14, 277n
Prabhat, Anjali, 99
predators, 6, 33, 36, 46, 104, 105,
    107–8, 111, 140–41
pregnancy
    fetus in, 166, 188
    mother's gait and posture during,
        190–91
    pelvic organ prolapse and, 193–94
    varicose veins and, 239

primates, 32, 35, 76, 107, 186, 282n
   birth in, 182
   gestational lengths in, 187–88
   young of, 164–65
   see also apes; monkeys
Principles of Geology (Lyell), 92
prosthetics, 244, 245, 264, 324n
psychopaths, 200–201

Raichlen, Dave, 213
Ratey, John, 231
Reed, Pam, 196
reptiles, 17–18, 23
rhinoceros, 131–32, 296n
Ridges, Phil, 33
Rising Star Cave, 156–58
Rizzo, Juliette, 216
Robinson, J. T., 120, 287n
robots, 236–37, 317n, 324n
Rosenberg, Karen, 183, 184, 262, 263
Rousseau, Jean-Jacques, 224, 321n
Rubin, Shirley, 97
Rudapithecus hungaricus, 82, 86
running, 6, 26, 35, 141, 245
   in athletic events, 194–96, 215
   of birds, 25
   shoes and, 249–50, 252
   Tarahumara and, 249–50
Ryan, Tim, 210

Sagan, Carl, 267–68, 325n
Sahelanthropus tchadensis ("Toumaï"),
   61–65, 68, 83–84, 86, 100, 109,
   287n
Salopek, Paul, 3
Sapiens (Harari), 187
Sapolsky, Robert, 261–62, 284n
Saptomo, Wahyu, 152
Savvides, Marios, 203
Science, 68
Scientific American, 181, 186
scoliosis, 241, 257
Senut, Brigitte, 59–60, 63–64, 286n
Shakespeare, William, 199
Shapiro, Liza, 190–91
shinbone (tibia), 80, 242, 243
Shipman, Pat, 135
shoes, 150–52, 249–51
SIDS, 168
Sima de los Huesos, 260
sinus cavities, 239–40
sloth, 30
Smith, Tanya, 111, 294n

Solnit, Rebecca, 196–97, 315n
species and genus, 10–11, 276n
spine and vertebrae, 51–52, 63, 81,
   190–91, 240–41, 257, 258, 284n
sports and athletics
   athletic ability, 171–72, 194–96
   injuries in, 308n
   Olympic Games, 26, 35, 195, 217, 244,
      246
   running events, 194–96, 215
Stern, Michael, 44
Sthenurus stirlingi, 29–30
stroke, 219
Strug, Kerri, 246
subtalar joint, 245, 246
Suwa, Gen, 67
Swift, Jonathan, 224
swimming, 14–15, 35
Switzer, Kathy, 194

tails, 26–27, 238
Tanner, Nancy, 45–46
Tanzania, 135
   Laetoli, see Laetoli
   Olduvai Gorge, 90, 91, 94, 100–102, 131
Tao Te Ching (Lao Tzu), 163
Tarahumara, 249–50
tarsometatarsus, 245
Tarus, Benyamin, 152–54
Taung child, 9–12, 14, 52, 119, 275n
teeth, 130
   canine, 8, 9, 45, 46, 63, 109
Tempest, The (Shakespeare), 199
testes, 236
thinking, and walking, 221–34
   brain connectivity and, 226
   cognitive decline and, 226–27, 229–31
   creativity and, 225–26
   depression and, 231–33
Thoreau, Henry David, 203, 221, 224
Thorpe, Susannah, 85
thumb, 85
Tianyuan Cave, 151
tibia, 80, 242, 243
Tobias, Phillip, 13, 276n
toes, 48, 53–54, 61, 69–71, 74, 75, 85,
   86, 102, 151, 248, 249
Tokei, Yamasa, 217–18
Tolkien, J. R. R., 144, 145, 154
tools, 8, 12–14, 100–101, 109, 110,
   130, 131–32, 138, 146, 148–49,
   152, 186–87
   animals' use of, 8, 103

tools (*continued*)
    bipedalism and, 7, 8, 12–14, 36, 44,
        101–4, 111–12
    weapons, 7, 12, 14, 36, 44
Tracks (Davidson), 225
Trevathan, Wenda, 184, 193, 262
Trevelyan, George Macaulay, 209
Triassic, 20, 22
Trinkaus, Erik, 151
Tsimané people, 213
Tsoukalos, Giorgio, 39
Tucker, Steve, 156
tumor necrosis factor (TNF), 215, 216
Tuttle, Russell, 32, 280n
2001: A Space Odyssey, 7, 13, 43
Tyrannosaurus rex, 20, 23–24, 26, 245,
    278n, 279n

Uganda, 170
uniformitarianism, 92–93
United Nations, 184–85
upright posture, 27, 30, 33–34, 42, 73
upright walking, *see* bipedalism
uric acid and uricase, 75–76, 289n
Ussher, James, 55

van Gogh, Vincent, 163–64
varicose veins, 239
Vekua, Abesalom, 131–32
vertebrae, *see* spine and vertebrae
violence, 12–13, 258–62, 265
    in chimpanzees, 258–59, 321n
    warfare, 322n
    weapons, 7, 12, 14, 36, 44
Vitruvian Man, 235
vocalizations, 141–42, 298n
volcanic eruptions, 55–56, 93, 99, 100,
    129, 150, 245, 292n
Vonn, Lindsey, 243
Vonnegut, Kurt, 219
Voss, Michelle, 226

Walker, Alan, 135
walking
    children's learning of, 163–78
    fetal movements and, 166
    gait and posture differences in, 199–207
    in groups, 191–92
    health benefits of, *see* health benefits of
        walking
    hyperpronation and, 122–25
    number of steps taken during a lifetime, 4
    as social phenomenon, 203–4

    speeds of, 204–6
    synchronized, 204
    thinking and, *see* thinking, and walking
    in the woods, 232–33
    words for, 4
    see also bipedalism
"Walking" (Thoreau), 221, 224
Walking Life, A (Malchik), 173
Wallace, Alfred Russel, 139
Wallace, Ian, 250
Wallace's Line, 139, 152
WALL-E, 219
Wall-Scheffler, Cara, 45, 189–92, 206
Wanderlust (Solnit), 196–97, 315n
Ward, Carol, 82–83
warfare, 322n
Warrener, Anna, 188–90
Washburn, Barbara, 186
Washburn, Brad, 185–86
Washburn, Sherwood (Sherry),
    185–87, 189, 190, 194
water, 41, 42, 115
Watts, David, 247
weapons, 7, 12, 14, 36, 44
weight loss, 214, 313n
Western, David (Jonah), 92
Wheeler, Peter, 42, 140
Whitcome, Katherine, 190–92
White, Tim, 61, 67–71, 73–74, 126,
    288n
Wisdom of the Bones, The (Walker and
    Shipman), 135
WoldeGabriel, Giday, 69
Wonderwerk, 140
Wong, Yan, 35
Woolf, Virginia, 225
Woranso-Mille, 126–28, 257
Wordsworth, William, 203, 224, 315n
World Health Organization, 167
Wrangham, Richard, 140, 258, 259,
    321n
Wu, Xiu-Jie, 260–61

Zallinger, Rudolph, 72, 73, 288n
Zanno, Lindsay, 20–21, 23, 24
Zelazo, Philip Roman, 166–67
Zeno's paradoxes, 316n
Zhang, Amey Y., 123–24
Zhu, Zhaoyu, 132
Zihlman, Adrienne, 45–46
Zipfel, Bernhard, 9, 70, 117–19, 121,
    123, 294n
Zivotofsky, Ari, 204